Mastering the Middle School TAKS in Science

MARK JARRETT

STUART ZIMMER

JAMES KILLORAN

JARRETT PUBLISHING COMPANY

East Coast Office
Post Office Box 1460
Ronkonkoma, NY 11779
631-981-4248

West Coast Office
10 Folin Lane
Lafayette, CA 94549
925-906-9742

1-800-859-7679   Fax: 631-588-4722
www.jarrettpub.com

This book includes material from many different sources. Occasionally it is not possible to determine if a particular source is copyrighted, and if so, who is the copyright owner. If there has been a copyright infringement with any material produced in this book, it has been unintentional. We extend our sincerest apologies and would be happy to make immediate and appropriate restitution upon proof of copyright ownership.

Copyright 2005 by Jarrett Publishing Company

Cover photo: Superstock, Inc.

All rights reserved. No part of this book may be reproduced in any form or by any means, including electronic, photographic, mechanical, or by any device for storage and retrieval of information, without the express written permission of the publisher. Requests for permission to copy any part of this book should be mailed to:

Jarrett Publishing Company
Post Office Box 1460
Ronkonkoma, New York 11779

ISBN 1-882422-86-4
Printed in the United States of America
First Edition
10 9 8 7 6 5 4 3 2 1   07 06 05

ACKNOWLEDGMENTS

The authors would like to thank the following Texas educators who helped review the manuscript. Their collective comments, suggestions, and recommendations have proved invaluable in preparing this book.

Charlotte Burns
Secondary Science Coordinator
Cypress-Fairbanks I.S.D.
Houston, Texas

Lisa Soll
Science Curriculum Specialist
San Antonio I.S.D.
San Antonio, Texas

Carole W. Henry
President of the Integrated
Science Educators in Texas
Assistant Curriculum Facilitator,
Southwest I.S.D.
San Antonio, Texas

Tiffany Huitt
Middle School Lead Science Teacher
Dallas I.S.D.
Dallas, Texas

Dee Wallace
Science Chair,
Sharpstown Middle School
Houston I.S.D.
Houston, Texas

Cheryl Willis
Science Chair,
Holland Middle School
Houston I.S.D.
Houston, Texas

Layout, graphics and typesetting: Burmar Technical Corporation, Albertson, NY.

This book is dedicated…

to my wife Goska Jarrett, and my children
Alex and Julia — *Mark Jarrett*

to my wife Joan, my children Todd and Ronald, and
my grandchildren Katie and Jared — *Stuart Zimmer*

to my wife Donna, my children Christian, Carrie, and Jesse, and
my grandchildren Aiden, Christian, and Olivia — *James Killoran*

TABLE OF CONTENTS

UNIT 1: INTRODUCTION

Introduction 1
Chapter 1: How to Answer Multiple-Choice Questions 2
Chapter 2: Interpreting Different Types of Data 10

UNIT 2: THE NATURE OF SCIENCE AND SCIENTIFIC INVESTIGATION

Chapter 3: The Process of Scientific Investigation 19
Chapter 4: The Growth of Scientific Understanding 38

UNIT 3: THE STRUCTURE AND PROPERTIES OF MATTER

Chapter 5: The Structure of Matter 49
Chapter 6: The Properties of Matter 62

UNIT 4 : THE NATURE OF MOTION, FORCE AND ENERGY

Chapter 7: Force and Motion 77
Chapter 8: The Nature of Energy 90

UNIT 5: LIFE SCIENCES

Chapter 9: The Cell 104
Chapter 10: Organisms and Organ Systems 114
Chapter 11: Heredity and Adaptation 127
Chapter 12: Ecosystems 138

UNIT 6: EARTH AND SPACE SYSTEMS

Chapter 13: The Universe 151
Chapter 14: Planet Earth: Cycles, Systems and Interactions 160

UNIT 7: A PRACTICE TAKS IN SCIENCE

Chapter 15: A Practice Middle School TAKS in Science 178
Glossary 195
Middle School TAKS Objectives in Science 199
Index 203

UNIT 1

INTRODUCTION

Everyone wants to get a high score on the **Middle School TAKS in Science**. Unfortunately, just wanting a high score is not enough. You will really have to work at it. With this book as your guide, you should be much better prepared for the test — and even enjoy studying for it. This book provides a complete review or "refresher" of the knowledge and skills you will need to do your best on the **Middle School TAKS in Science**.

UNIT 1: TEST-TAKING SKILLS

The first three chapters of this book explain how to answer multiple-choice questions. These chapters review every major type of question you will find on the test.

UNITS 2 – 6: CONTENT REVIEW

The second part of the book consists of four units surveying what you have learned in middle school science. Each chapter —

★ opens with a list of *Major Ideas* highlighting the most important information in that chapter

★ is divided into small, but comprehensive, sections to help you understand a major field of science

★ includes *Applying What You Have* activities to help you think about and apply what you have just learned

★ includes **Study Cards** to help you review the most important *terms*, *concepts*, and *relationships* in the chapter.

★ finishes with a *What You Should Know* box, summarizing major ideas and facts in the chapter that are often the focus of TAKS questions

★ provides TAKS-style practice questions; each question is identified by its TAKS objective and a number indicating its grade level knowledge and skill statement.

You will also find a checklist of TAKS objectives at the end of each unit. Make sure you have mastered each objective before moving to the next unit.

UNIT 7: A FINAL PRACTICE TEST

The final part of the book consists of a *complete* practice test, just like the actual **Middle School TAKS in Science**.

CHAPTER 1

HOW TO ANSWER MULTIPLE-CHOICE QUESTIONS

All of the questions on the **Middle School TAKS in Science** will be multiple-choice questions. They will ask you to select the *best answer* from four choices. Let's look at each of the different types of questions on the TAKS, based on what they ask you do.

COMPREHENSION QUESTIONS

Comprehension questions will test your basic understanding of scientific concepts, facts, and relationships. These types of questions will usually ask you to do **one** of the following:

- Identify
- Describe
- Provide Examples

IDENTIFY

Identify questions test your ability to recall scientific information or to understand data provided in the question.

> **Note:** The examples below only show the beginning of each question. On the actual TAKS, these questions would be followed by a series of four answer choices, which are not included here.

★ Which process allows plants to store energy from sunlight?

★ Which subatomic particle has a positive electrical charge?

DESCRIBE

Describe questions ask you to tell about the features or characteristics of something. You could be asked *what* something consists of, *how* it looks, or *how* it works. Examine the following questions, which test your ability to *describe*:

★ Which is the best **description** of the human digestive system?

★ Which statement best **describes** the characteristics of a star?

PROVIDE EXAMPLES

Some questions on the TAKS test your understanding of **generalizations** — general statements that identify a pattern or rule that applies to many different examples. These questions will usually first state a scientific principle or concept. The question will then ask you to select an *example* from the answer choices that illustrates this generalization. Look at the following question, which tests your ability to *provide an example*:

> 1 **Which of the following is an example of a chemical change?**
>
>   A Ice melting
>   B Sugar dissolving
>   C Wood burning
>   D Rock splintering

This question tests your understanding of a general concept — *chemical change*. The answer choices provide specific examples of materials that are changing. Only one of the answer choices, however, provides an example of a *chemical* change. Can you determine which one of the choices it is?

ANSWERING COMPREHENSION QUESTIONS

To answer *comprehension questions*, try using the three-step "**E**-**R**-**A**" Approach:

[EXAMINE the Question]   [RECALL What You Know]   [APPLY What You Know]

For example, examine the question below:

> 2 **Green plants transform light energy into chemical energy through the process known as —**
>
>   F fermentation
>   G respiration
>   H photosynthesis
>   J excretion

UNLOCKING THE ANSWER

Step 1: EXAMINE the Question
Carefully read the question. Make sure you understand all the information it includes. Look at the answer choices. Eliminate choices that you know are wrong.

Question 2 asks about the process that transforms light energy into chemical energy. The question describes the process, while the choices give different names for this process. Your task is to select the correct name.

Step 2: RECALL What You Know
Now identify the subject that the question asks about. Take a moment to think about what you know about that subject. Mentally review the most important *concepts*, *facts*, and *relationships* that you can remember.

What do you remember about this subject? Recall that plants use sunlight to turn carbon dioxide and water into glucose and other carbohydrates. Energy from sunlight is stored in the chemical bonds of these organic compounds.

Step 3: APPLY What You Know to Answer the Question
Apply the information that you remember to select the correct answer.

*To answer this question, recall that **photosynthesis** is the name given to the process of turning sunlight into chemical energy. The answer is choice H.*

ANALYSIS QUESTIONS

Analysis questions test your ability to think like a scientist. To *analyze* something means to break it up into its various parts to understand it better. *Analysis questions* might ask you to *sequence*, *explain*, or *compare*.

SEQUENCE

Sequence questions focus on the order in which things happen. They may ask you to identify the correct order of steps in a process, or the order in which a series of events occurred.

Look at the following questions which test your knowledge of *sequence*:

★ What **step is missing** from this diagram of the water cycle?

★ Which diagram shows how energy **flows** through an ecosystem?

EXPLAIN

Explain questions ask you about *cause-and-effect* relationships. You could be asked to identify the *causes* of something — what made it happen — or the *effects* of an event or process.

CAUSE Someone turned on the switch. ➤➤➤➤➤ **EFFECT** The light went on.

Questions Asking for Causes

★ Which best **explains why** dinosaurs became extinct?

★ **Why would** a moving object change its speed?

Questions Asking for Effects

★ Which of the following is an **effect** of soil erosion?

★ Which of the following is a direct **result** of tectonic plate movements?

An *explain question* might also ask for the best scientific explanation or evidence:

★ Which **evidence** best **supports** the theory that Earth's crust consists of giant shifting plates?

COMPARE

Compare questions will ask you to identify how two or more items are alike or different. The following sample questions test your ability to *compare*:

★ How does a neutron **differ** from a proton?

★ How is a plant cell **different** from an animal cell?

ANSWERING ANALYSIS QUESTIONS

To answer questions that ask you to *sequence*, *explain*, or *compare* again use the "**E**xamine – **R**eview – **A**pply" Approach. Look at the following sample question:

3 A scientist plants a row of trees on a grassy plain as a method of soil conservation. How do the trees protect the soil from erosion?

   A   The leaves of the trees convert sunlight into chemical energy.
   B   Trees attract birds and insects, which make their home in the tree branches.
   C   The roots of the trees help to hold the soil together.
   D   The trees shade the soil from the harmful rays of the sun.

Let's see how the **"E-R-A" Approach** helps you find the correct answer:

UNLOCKING THE ANSWER

Step 1: EXAMINE the Question
Carefully read the question. Examine any data or information it provides. Make sure you understand what the question is looking for. Also look at the answer choices.

In question 3, a scientist has planted trees. Your task is to explain why a scientist would do this to protect soil from erosion. The answer choices provide possible ways in which the trees help to protect the soil.

Step 2: RECALL What You Know
Now identify the subject of the question. Take a moment to think about what you can remember about that subject. Mentally review the most important facts and concepts that you can recall.

This question is about soil erosion. Think about what you can recall about soil erosion and its causes. You may remember that wind and rainfall often cause soil to blow or wash away.

Step 3: APPLY What You Know to Answer the Question
Apply the information that you remember to select the correct answer.

To answer this question, you must apply what you remember about the causes of soil erosion. Although the leaves of the tree convert sunlight through photosynthesis (Choice A), this does not help reduce soil erosion. The same is true for choices B and D. However, Choice C explains how trees act to reduce soil erosion. Tree roots extend deep into the soil, holding it together against the forces of wind and rain.

INFERENCE QUESTIONS

To *infer* means to use your own logical thinking to draw conclusions about information or to apply it in a new way. *Inference questions* on the **Middle School TAKS in Science** will ask you to:

- Draw Conclusions
- Make Predictions
- Apply Scientific Knowledge to New Situations

DRAW CONCLUSIONS

Conclusion questions provide you with data or observations and ask you to make a generalization about them. Ask yourself: What does this evidence show? You can think of *conclusion questions* as being the opposite of questions that ask you to provide examples:

> ★ *Example questions* state a general principle and ask you to identify a specific example that illustrates it.
>
> ★ *Conclusion questions* give a series of specific examples or details and ask you to identify a general principle that those examples show.

Examine the following *conclusion questions*:

 *What **conclusion** can be drawn from the results of this experiment?*

 *What **conclusion** about the physical properties of metals can be drawn from these examples?*

MAKE PREDICTIONS

Prediction questions test your ability to tell the future. These questions often ask you to apply your knowledge of scientific cause-and-effect relationships to new situations. If one factor, or *variable*, changes, how will other factors be affected? Look at the following example of a *prediction question*:

★ *Which gas in the atmosphere would increase if a large number of new trees were planted?*

APLYING SCIENTIFIC KNOWLEDGE TO NEW SITUATIONS

Some *inference questions* will ask you to apply your scientific knowledge to new situations — either in the laboratory or in the "real world" outside the classroom. These questions might ask you the best way to reach a desired outcome.

★ *Iron filings are mixed together with yellow sulfur powder.* **Which would be the best way** *to separate the iron filings from the powder?*

This type of question might also ask you to select the best laboratory procedures to test a hypothesis:

★ *What would be the best procedure for testing the hypothesis that variations in the amount of sunlight affect plant growth?*

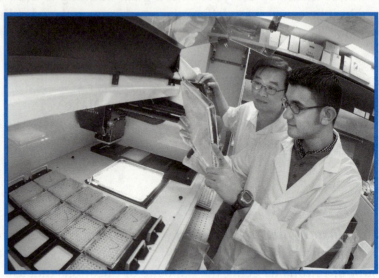

Scientists often apply scientific knowledge to new situations

You could be asked to apply your scientific knowledge to make a calculation:

★ *What is the density of an object with a mass of 20 grams and a volume of 2 cm^3?*

ANSWERING INFERENCE QUESTIONS

To answer any question asking you to *draw conclusions, make predictions*, or *apply scientific concepts to new situations*, you should again use the **"E-R-A" Approach**. For example, look at the sample question below. Notice how it asks you to apply your scientific knowledge to a "real world" situation.

---

4 A world health organization is concerned about the effects of increasing amounts of carbon dioxide in Earth's atmosphere. What step can it suggest to reduce these amounts?

  F Discourage the use of chemical pesticides
  G Encourage the reforestation of tropical rainforests
  H Encourage the planting of farm crops in place of forests
  J Discourage the use of nuclear energy as a source of power

CHAPTER 1: HOW TO ANSWER MULTIPLE-CHOICE QUESTIONS

To answer *inference questions*, take the following steps:

UNLOCKING THE ANSWER

Step 1: EXAMINE the Question

First, read the question carefully, examining any data it may include. Then focus on what the question asks for.

In this case, the question asks about steps that can be taken to protect the atmosphere from increasing levels of carbon dioxide.

Step 2: RECALL What You Know

This kind of question asks you to apply your scientific knowledge to new circumstances. You need to draw a general conclusion from specific scientific evidence, or to identify which scientific concepts best apply to the situation. Take a moment to review the appropriate facts, concepts, and relationships in your mind.

To answer question 5, you might recall that plants absorb carbon dioxide in the process of photosynthesis.

Step 3: APPLY What You Know to Answer the Question

Now you are ready to apply what you recalled to the question. Think about how the scientific concepts you recalled can be used to resolve the problem posed by the question.

In question 4, which answer choice would help reduce carbon dioxide in Earth's atmosphere? Because trees absorb more carbon dioxide than farm crops do, many environmentalists are troubled by the destruction of the world's tropical rainforests. Regrowing rainforests would help to absorb more carbon dioxide. Therefore, the correct answer choice is G.

*The other answer choices are clearly wrong. The use of fewer pesticides (**Choice F**) would not affect the amount of carbon dioxide in Earth's atmosphere. Nuclear energy is used as an alternative to the burning of fossil fuels (oil, coal, and natural gas). Using less nuclear energy would probably increase the use of fossil fuels, producing more, not less, carbon dioxide.*

CHAPTER 2

INTERPRETING DIFFERENT TYPES OF DATA

Many of the questions on the **Middle School TAKS in Science** will include observations or data that you have to interpret. These include:

In this chapter, you will learn how to interpret these forms of data on the TAKS. Remember to use the **"E-R-A Approach"** — *Examine*, *Recall*, and *Apply* — when answering any question that includes data:

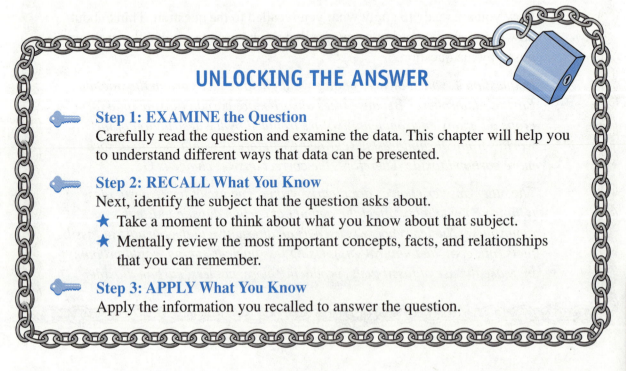

UNLOCKING THE ANSWER

Step 1: EXAMINE the Question
Carefully read the question and examine the data. This chapter will help you to understand different ways that data can be presented.

Step 2: RECALL What You Know
Next, identify the subject that the question asks about.
★ Take a moment to think about what you know about that subject.
★ Mentally review the most important concepts, facts, and relationships that you can remember.

Step 3: APPLY What You Know
Apply the information you recalled to answer the question.

CHAPTER 2: INTERPRETING DIFFERENT TYPES OF DATA

TABLES

A **table** is used to organize information. Tables list information in columns and rows. To interpret a table, pay close attention to the headings of the columns and rows, and to the units of measurement that are used.

Variables. A scientific table often shows the relationship between two things. Because both items can change, or *vary*, they are known as **variables**. Scientists usually have control over the first variable — known as the **independent variable**. They change this first variable to see what effect this has on the second variable — known as the **dependent variable**.

At the same time, scientists try to hold all other factors constant. This allows them to really see what impact the changes in the independent variable have on the dependent variable. Let's see how this works by examining the following table and information:

A group of scientists has been studying the amount of oxygen that can be dissolved in water at different temperatures. They have gradually heated the water and measured how much oxygen can be dissolved in it at 5° Celsius intervals.

AMOUNT OF OXYGEN THAT CAN BE DISSOLVED IN WATER

Water Temperature (C°)	Oxygen Content* (*parts per million*)
10	11.29
15	10.10
20	9.11
25	8.27
30	7.56

***Note:** Parts per million* means that in a given amount of water, one-millionth of it is dissolved oxygen. For example, at a water temperature of 10° C, you will find up to 11.29 *parts* of oxygen can be dissolved in every *million parts* of water.

In this experiment, scientists measured the amount of oxygen that could be dissolved in fresh water at various temperatures. They recorded their results in the table.

★ The first column shows the **independent variable** — the temperature of the water in degrees Celsius (C°). This variable was controlled by the scientists.

★ The second column shows the **dependent variable** — the amount of oxygen that can be dissolved in water at that temperature, in parts per million.

1 According to the table, how much oxygen can be dissolved in water at 20°C?

    A 9.11 parts per million   C 9.11 millimeters
    B 8.27 cubic centimeters   D 8.27 mL

12 MASTERING THE MIDDLE SCHOOL TAKS IN SCIENCE

This question tests your understanding of the table and its units of measurement. At 20°C, the amount shown on the table in the second column is "9.11". According to the heading at the top of the second column, the number 9.11 refers to "parts per million." This means that in one million parts of water at 20°C, up to 9.11 parts can be dissolved in oxygen — **Choice A**.

2. According to the data in the table, what happens to the amount of oxygen that can be dissolved in water as its temperature rises?

   F   Increases
   G   Decreases
   H   Remains the same
   J   Decreases, then increases

♦ Examine the Question
♦ Recall What You Know
♦ Apply What You Know

This question asks you to *analyze* the information in the table. To answer this question correctly, study the table and try to determine the relationship between the **independent variable** (*temperature*) and the **dependent variable** (*oxygen content*). The clue to answering this question correctly is to find out what happens to the temperature as you move down the table. What happens to the amount of dissolved oxygen as the temperature rises?

LINE GRAPHS

A *line graph* shows a series of points on a grid connected by a line. The title tells what the graph shows. Each point on the grid represents a specific quantity. The purpose of a line graph is to show how two or more variables are related.

Usually, a line graph shows how changes in one variable will lead to changes in the other variable. The **X axis** usually represents the *independent variable*, and the **Y axis** represents the *dependent variable*. The X axis is the bottom line of the graph. As you move to the right of the graph, the value of X on this line increases. The left side of the graph is the Y axis. As you move up this line, the value of Y increases.

To remember the differences between the X and Y axes, you might think of the phrase "Dry Mix":

Dependent	**R**esponding	**Y** Axis
Manipulated	**I**ndependent	**X** Axis

The graph below shows information about ocean temperatures. Here, scientists have measured the temperature of the ocean at different depths. Which is the **independent variable**? Which is the **dependent variable**? As you can see from this example, a line graph often shows patterns better than a table does.

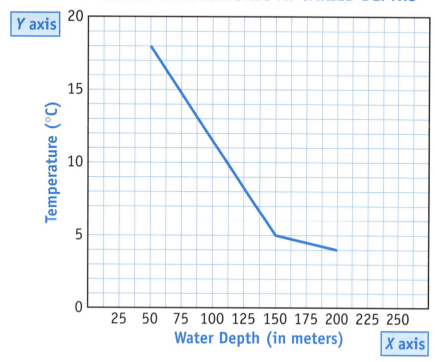

3. Based on the graph, what happens to the water temperature as the ocean depth increases?

   A   Remains constant
   B   Decreases
   C   Increases
   D   Increases, then decreases

HINT

As you have just read, each **axis** represents a different variable. The *horizontal (or X)* axis shows the depth of the water in meters. The *vertical (or Y)* axis shows the temperature of the ocean at that depth. Each point on the graph represents the temperature (*the dependent variable*) at a particular water depth (*the independent variable*). Now that you understand this, can you determine what happens to the temperature as you go deeper into the ocean?

APPLYING WHAT YOU HAVE LEARNED

✦ Since you now know how to read a line graph, let's see if you can also make one. Review the table on page 11 and turn its information into a line graph.

BAR GRAPHS

A **bar graph** is made up of parallel bars of different lengths. It is used to compare different items or one item over time. Each bar represents a quantity. Either each bar is labeled, or a key explains what each bar represents. The Y axis often indicates what quantities the bars show.

Use the bar graph below to answer the question.

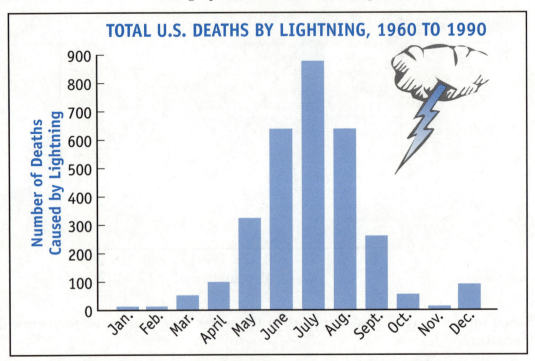

4 What conclusion can be drawn from the information in the bar graph?

    F Most deaths from lightning occur along the eastern coastline.

    G The average number of deaths each year is about 900.

    H The greatest number of deaths generally occur in the summer months.

    J Most deaths from lightning occur as a result of hurricanes.

♦ **Examine the Question**
♦ **Recall What You Know**
♦ **Apply What You Know**

This question tests your ability to draw conclusions from a bar graph. Here, each bar shows the number of deaths from lightning in a particular month over a 30-year period. Thus, 900 people died from lightning strikes in July between 1960 and 1990. To answer the question correctly, you should use the information in the bar graph to test each answer choice. For example, the graph does not tell whether most deaths from lightning occurred on the eastern coastline. Answer Choice F is therefore wrong.

CHAPTER 2: INTERPRETING DIFFERENT TYPES OF DATA   15

FLOW CHARTS

A **flow chart** is a special type of diagram. It shows a series of steps in a process. Each step is placed in some geometric shape or represented by a picture. The shapes or pictures are connected by arrows or lines to indicate the order of steps in the process. Sometimes the process may move forward in more than one way. Arrows will indicate when a step may lead to different choices or outcomes.

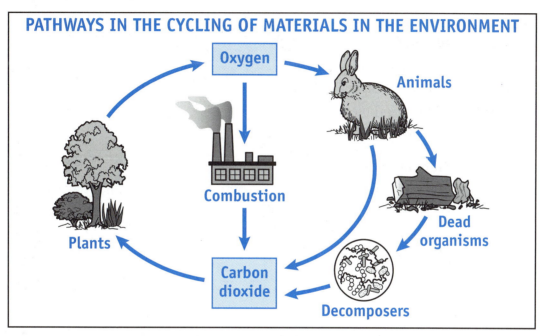

5 Based on the flow chart above, which of the following is a source of oxygen for the atmosphere?

   A   combustion
   B   plants
   C   decomposers
   D   animals

♦ Examine the Question
♦ Recall What You Know
♦ Apply What You Know

To answer this question, you have to understand how a flow chart works. The arrows indicate how organisms produce and consume two gases — oxygen and carbon dioxide. Using the flow chart, notice that:

♦ Plants, such as trees, consume carbon dioxide and produce oxygen.
♦ Animals and combustion processes consume oxygen and produce carbon dioxide.
♦ Decomposers produce carbon dioxide when they break down dead organisms.

Based on these facts, which answer is correct?

MAPS

A **map** is a diagram representing a place. It shows where objects are located. Maps may be used to show the location of the stars or planets, or to depict characteristics of Earth's surface. The **legend** of the map explains symbols used on the map. The **direction indicator** shows directions on the map. The **scale** tells what distances on the map represent in the actual area.

6   Which conclusion can be drawn from the map?

   F   Illinois produces more air pollution than other states.

   G   The air pollution problem in Baltimore is increased by the addition of pollution from other areas.

   H   There is no air pollution south of Virginia.

   J   Air pollution problems in Virginia clear up quickly as the air moves toward the sea.

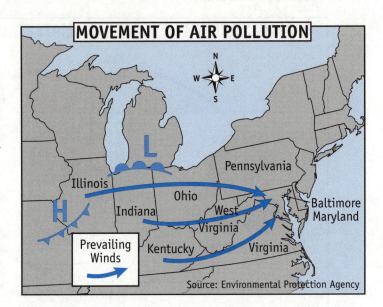

To answer question 6, you have to study the information on the map. The arrows show that winds push air pollution from the Midwest across the eastern United States. From the information, we cannot conclude that choices F, H or J are correct. We can determine that these winds converge on Baltimore.

WEATHER MAPS

Weather maps show the conditions in Earth's atmosphere. A weather map can show temperatures, rain and snow, winds, humidity and air pressure.

★ **Air Pressure.** Air pressure is the force with which the air presses on the ground. On a weather map, areas of dense air exert high pressure and signal the approach or continuation of fair weather. It is shown with an "H." See the "H" on the map above. Low pressure areas are shown with an "L," as indicated on the map above. Falling pressure usually indicates a storm.

★ **Warm and Cold Fronts.** A **front** is a place where two different air masses meet. A cold front is a mass of cooler air, shown as a line with triangles. The triangles show the direction the front is moving. A warm front, a mass of warmer air, is shown as a line with semicircles.

7 Based on the map on page 16, the weather in southern Illinois will most likely —

    A become cooler   C become wetter

    B become warmer   D remain the same

CONTOUR MAPS

On the **Middle School TAKS in Science**, you may be asked to read a contour map. A **contour map** is a special kind of map that shows the height or elevation of land areas. Look at the following picture of an island. This island is mountainous. Imagine that someone has painted lines on the mountains, and that each line is 500 feet higher than the line before it.

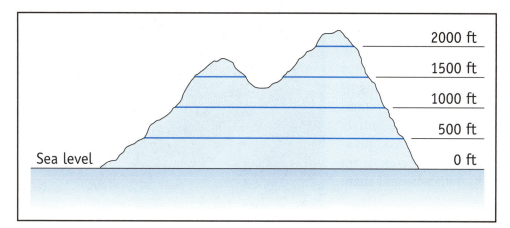

These lines are called **contour lines**. They show the shape of the island. If you took a helicopter ride over the island, you could imagine its contour lines. They would make up the contour map on the right:

The distance between the contour lines is known as the **contour interval**. In the case of this map, the contour interval is 500 feet. Notice that on the right side of the island, the slope rises more steeply. Thus on the contour map, the contour lines on that side are closer together.

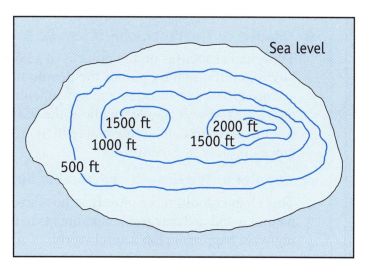

APPLYING WHAT YOU HAVE LEARNED

♦ Make an imaginary contour map of your backyard, school yard or community. On your contour map, show the main features of the place you choose and draw contour lines to show where the land rises and falls.

UNIT 2: THE NATURE OF SCIENCE AND SCIENTIFIC INVESTIGATION

Science is a systematic way of investigating and explaining the natural world. The **Middle School TAKS in Science** will test your knowledge of several fields of science.

The first unit of this book reviews common principles that apply to all the fields of science. To try to understand our natural world, scientists make observations, ask questions, develop theories, form and test hypotheses, and share ideas.

★ **Chapter 3: The Process of Scientific Investigation**

This chapter focuses on the process of scientific investigation. You will learn how scientific inquiry guides the development of scientific knowledge, how science is based on theories, and how scientists test their theories through investigation. You will also learn about the variety of tools and methods used by scientists when conducting a scientific investigation, and about the importance of safety in experiments.

★ **Chapter 4: The Growth of Scientific Understanding**

This chapter looks more closely at how scientists use their observations of the natural world and data from experiments to develop and revise theories explaining what happens in the natural world.

CHAPTER 3

THE PROCESS OF SCIENTIFIC INVESTIGATION

In this chapter, you will learn how scientists think and investigate.

MAJOR IDEAS

A. **Scientific inquiry** asks questions about the natural world in a systematic way.

B. A **scientific theory** attempts to explain data received from observing nature and conducting experiments. Models and mathematics help scientists develop theories.

C. The following steps are often used to conduct experiments. Their order may vary:

★ Observe the world and develop a hypothesis.

★ Design an experiment to test your hypothesis. Identify the independent and dependent variable in your experimental design.

★ Select the best equipment and technology to conduct your experiment.

★ Take safety into account in all field and laboratory investigations.

★ Conduct your experiment, organize your results and relate them to your hypothesis.

★ Communicate your results and conclusions to others.

METHODS OF SCIENTIFIC INQUIRY

Scientific inquiry begins when a scientist observes the natural world. The scientist then asks questions about those observations. For example, a scientist may see an apple fall to the ground and ask: "Why did it fall? How fast did it fall? What factors affected its speed and direction?"

Scientists then try to find answers to their questions, to explain what they see. To find these answers, scientists carry out research, build models, make new **observations** and conduct experiments.

Theories. To explain their research findings, observations and experimental results, scientists develop theories. **Theories** apply general principles to explain large amounts of information and observations. For example:

★ The Big Bang theory says the universe began with a single explosion.

★ Cell theory says every living thing is made of cells.

Scientists continually test, retest, and revise their theories.

Models and their Limitations. Often a theory is based on a **model**. A model is a diagram or object that represents something else and demonstrates how it works. For example, an architect may build a miniature model of a house to see how it will look and to check its proportions before constructing the house itself. Similarly, a chemist may create a plastic model of a molecule to understand it better.

How do investigations help to build scientific knowledge?

Models, however, can never be exactly the same as what they represent. They will always differ in some way in size, materials, speed of movement or some other factors. Because of these differences, models always have some limitations and can always be improved. The more closely a model resembles the real thing, the better it generally is.

Some questions on the TAKS may test your understanding of models and their limitations. Remember, every model is a simplified version of reality. Each model can be made more realistic and accurate by adding more details. The key to answering a *model* question correctly is to have a clear understanding of how closely the model resembles what it represents. *Model* questions on the TAKS will usually describe the model in an introductory sentence or two, or provide an illustration of the model. Most often, model questions will ask one of the following:

| How closely does it resemble what it represents? | What does the model show? | How might the model be improved? |

The Role of Mathematics. Many modern scientific theories are influenced by mathematics, which can be applied to anything that can be measured. Once scientists develop a model for how something works, they can apply mathematics to make predictions and to test hypotheses. Creating a mathematical equation helps scientists to represent a complex relationship precisely by using only a few symbols and numbers.

APPLYING WHAT YOU HAVE LEARNED

✦ Suppose you wanted to explain the solar system to your younger brother or sister.
  • How would you make a model of the solar system?
  • What would be the limitations of your model?

THE KEY STEPS OF SCIENTIFIC INVESTIGATION

Scientists use special methods to investigate the natural world. They observe nature and take careful measurements. They also try to control conditions either in the laboratory or in the field to test their theories and ideas. Such a test under controlled conditions is known as an **experiment**. Specific procedures vary from scientist to scientist and from experiment to experiment, but good scientific work always follows a logical, step-by-step approach.

While there is no single method of scientific investigation, many scientists use the following steps:

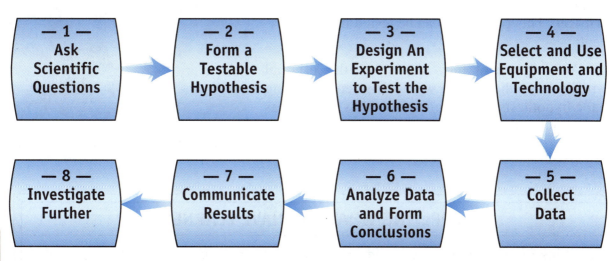

The order of these steps may vary. Sometimes, scientists may even go back to an earlier step in their investigation to change and improve their experimental design.

ASK SCIENTIFIC QUESTIONS

Based on their observations of nature and their own theories, scientists develop questions. Then they refine these questions for both field and laboratory investigations.

Questions that are vague or that ask for opinions cannot be answered by scientific investigation. Questions for investigation must be *factual* and *specific*. A good research question identifies *exactly* what will be identified in an experiment or field study. For example, the following question is *not* precise enough for a specific experiment:

What is the effect of fertilizers on plants?

Scientists will wonder which fertilizers, what kind of effect, and what kind of plants the question refers to. A better question for research would be: *How does nitrogen-based fertilizer affect the growth of bean plants*? This new question states what kind of fertilizer is being used, what effect is being observed, and what kind of plant is being studied.

FORM A TESTABLE HYPOTHESIS

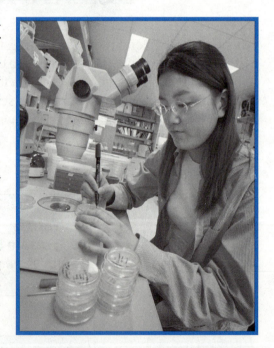

A **hypothesis** is an educated guess, often based on a theory, that attempts to answer the question under investigation. An ideal hypothesis can be **tested** in an experiment. For example, a scientist may make the hypothesis that *nitrogen-based fertilizers help bean plants to grow*.

Often a scientist turns the hypothesis into a **prediction**: *If bean plants are given nitrogen–based fertilizer, they will grow faster than plants without fertilizer*. This prediction can be tested in an experiment.

An experiment may show that the original hypothesis is either right or wrong. In science, proving that a hypothesis is wrong can be just as valuable as proving it is right.

APPLYING WHAT YOU HAVE LEARNED

✦ Think of an experiment you did in science class this year. What hypothesis did that experiment test?

✦ Why do you think it can be just as important to prove that a hypothesis is wrong as to prove it is correct?

DESIGN AN EXPERIMENT TO TEST YOUR HYPOTHESIS

An experiment creates **controlled conditions** to test the hypothesis. The **experimental design** is the road map of the experiment.

Experiments deal with independent and dependent variables.

Independent Variables. The researcher must have a clear understanding of the **variables** (*things that can change*) in the experiment. Some of these variables can be changed by the experimenter. For example, a researcher conducting a plant experiment can decide how much fertilizer to use, how much water to use, and how long to keep the plants in sunlight. Each of these is an example of an independent variable.

Dependent Variables. Other variables cannot be directly controlled by the experimenter. The height that each plant grows is an example of this second kind of variable, known as the **dependent variable**. What happens to the dependent variable is the **effect** of what the experimenter does to the independent variable. It is a **response** to the changes in the independent variable. Some scientists call this the "responding variable."

Controlling Variables. In a controlled experiment, the experimenter usually changes only *one independent variable* at a time. For example, in an experiment to test nitrogen-based fertilizer, a scientist will select one independent variable, such as how much fertilizer a plant receives. The researcher will change that variable and see how that change affects another variable, such as plant growth. Other variables will stay the same. The researcher will use only one kind of bean plant, with all plants being as close to the same size as possible. The experimenter will provide the same amount of water at the same time of day to all the plants. The plants will all be exposed to the same amount of sunlight. Everything will stay as close to the same as possible except the tested variable.

An Experiment has Different Kinds of Variables

★ **Independent Variable.** A variable an experimenter changes or manipulates to find out how this variable affects one or more other variables in the experiment.

★ **Dependent Variable.** A variable that changes in response to a change made to the independent variable in the experiment. Scientists usually measure this response.

★ **Variables Held Constant.** All the other variables in an experiment should be kept the same (*for example, the type of plant used and the amount of sunlight*).

Experimental Group vs. Control Group. Some plants, known as the experimental group, are given nitrogen-based fertilizer; a second group of plants, known as the **control group**, receives none. The scientist then can compare the results of the experimental group with the control group. The experimenter can be fairly certain that any significant differences in plant growth were probably caused by whether or not the plants received the fertilizer.

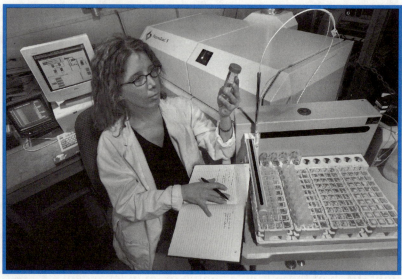

Comparing results between groups is essential to a scientific experiment.

The goal of an experiment is to obtain accurate and reliable results. By repeating the experiment several times in different *trials*, the experimenter can increase confidence in the results of the experiment.

APPLYING WHAT YOU HAVE LEARNED

✦ Why does a scientist generally change only one variable at a time in an experiment?

✦ Why does a scientist usually include both an experimental group and a control group when conducting an experiment?

USING EQUIPMENT AND TECHNOLOGY

Designing an experiment is a little like using a recipe. The experimenter must identify the required materials, equipment and technology. Then the experimenter must list the steps that need to be performed to conduct the experiment.

ELEMENTS OF A GOOD EXPERIMENTAL DESIGN

★ All the variables need to be identified, and only one independent variable is experimental.

★ All the necessary materials are listed, including their amounts.

★ Variables can be precisely measured with instruments.

★ There should be several trials.

STANDARD LABORATORY EQUIPMENT

The following are some of the standard types of laboratory and field equipment you should be familiar with before taking the **Middle School TAKS in Science**:

★ **Beaker.** A container used to hold liquids.

★ **Petri Dish.** A circular, covered glass dish used to hold biological samples.

★ **Meter Stick.** A ruler marked in centimeters and meters, used to measure length.

★ **Graduated Cylinder.** A glass cylinder marked in milliliters, used to measure the volume of liquids. When measuring the volume of a liquid, you must consider the meniscus.

Petri dishes

★ **Hot Plate.** An electrical plate used to heat substances in a laboratory investigation.

★ **Test Tube.** A glass tube used to hold or heat substances.

★ **Safety Goggles.** Protective plastic goggles worn during an experiment.

★ **Spring Scale.** An instrument that measures mass by seeing how much a hanging object pulls down a spring.

★ **Microscope.** A compound light microscope uses a series of lenses to magnify specimens placed on slides.

★ **Telescope.** An instrument that uses lenses and mirrors to magnify distant objects that appear small, such as planets or stars.

★ **Dissecting Equipment.** Equipment such as a scalpel, forceps, and dissecting probe used to cut open and examine a preserved specimen.

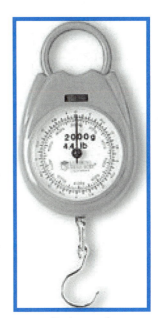
Spring scale

★ **Weather Instruments.**

  • **Thermometer.** An instrument that measures temperature. Scientists measure temperature in degrees Celsius (°C). This scale is based on the freezing and boiling points of water. The freezing point of water is 0°C, and the boiling point for water is 100°C.

  • **Barometer.** An instrument that measures atmospheric pressure. Changes in barometric pressure usually indicate a change in the weather.

  • **Humidity Meter (Sling Psychrometer).** An instrument that measures the amount of water vapor or moisture in the air.

★ **Water-Test Kit.** A kit with instruments, such as a pH meter, used to test water in field investigations.

★ **Timing Devices.** A clock, stopwatch, or other instrument used to measure time.

USING A BALANCE

A **balance** is an instrument used to measure mass. A simple balance is a bar with a pan on each end. The object to be measured is placed in one pan, and known units of mass are placed in the other. When the bar stays level, the masses in the pans are equal.

A **triple beam balance** has a single pan. Three scaled beams with "riders" are used to measure the mass of what is placed in the pan.

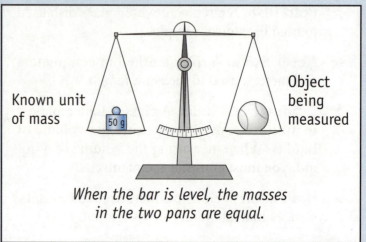

When the bar is level, the masses in the two pans are equal.

★ First, move the three riders to the left, so that the balance reads zero. Then place the object to be measured in the pan.

★ Move the 100 gram rider along the notches to the right until the indicator is just about to drop below the fixed mark.

★ Then do the same with the 10 gram rider.

★ Finally, move the 1-gram rider to the right until the indicator is lined up exactly with the fixed mark.

TRIPLE BEAM BALANCE

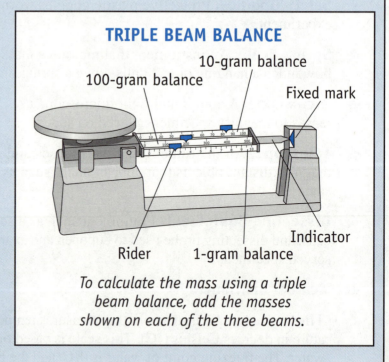

To calculate the mass using a triple beam balance, add the masses shown on each of the three beams.

SAFETY PRECAUTIONS

Attention to safety is essential during both field and laboratory investigations. Safety must be considered even before an experiment begins, and sufficient safety precautions should be included in the experimental design.

CHAPTER 3: THE PROCESS OF SCIENTIFIC INVESTIGATION

Important safety concerns include the following:

★ How will hazardous chemicals be used and do they produce dangerous fumes?

★ Are there any biological hazards, and how should they be handled?

★ Should the experimenters protect themselves with safety goggles and laboratory aprons?

★ Will open flames be used? Is a fire extinguisher nearby?

Students working together on a class experiment while employing safety practices.

Safety Equipment. Your classroom laboratory should have a fire extinguisher, fire blanket, first aid kit, safety goggles, and eye and face wash stations. You should know where these are located in the classroom.

Material Safety Data Sheets. Before using any chemicals, scientists consult the Material Safety Data Sheet (MSDS) for information about each chemical. The MSDS identifies potential hazards and the appropriate protective equipment that should be worn when handling that material. It also tells what to do in case of accidental contact or other emergency.

SOME COMMON LABORATORY SAFETY RULES

★ Follow your teacher's directions at all times.

★ Wear safety goggles, a laboratory apron, and gloves for most experiments.

★ Read all chemical labels and safety symbols.

★ Never heat liquids in closed containers. Always point the open end of test tube away from others when heating liquids or mixing chemicals.

★ Know where to wash your hands or eyes and where the fire extinguisher is located.

★ Clean up your work area and put materials away after you are done.

★ Wash hands with soap and water before and after all experiments.

APPLYING WHAT YOU HAVE LEARNED

✦ Describe one laboratory experiment you conducted.
  • What hypothesis did you test?
  • What laboratory equipment did you use?
  • What safety precautions did you follow?

COLLECT DATA

Information collected during an investigation is known as **data**. Data can be in the form of precise measurements or simple observations. A scientist uses a notebook as a diary or log during a laboratory experiment or field investigation. When something is done, observed, or measured, an entry is made in the notebook. All entries are dated.

Taking Accurate and Precise Measurements. The measurements recorded in the laboratory or field notebook come from using appropriate tools. Measurements are made using the international standards (*SI*) or **metric system**. For example, you would use a meter stick to measure the height of a bean plant in centimeters. The data you record cannot be any more precise than the instruments that you use can measure. All measurements should be taken several times and recorded with their units (*for example, centimeters*) in the log.

COMMON MEASUREMENTS USED IN EXPERIMENTS

Measurement	Length		Mass		Volume	
Basic Unit	meter	m	gram	g	liter	L
Smaller Unit	centimeter	cm	milligram	mg	milliliter	mL
Same or Larger Units	kilometer	km	kilogram	kg	cubic centimeter	cm³

★ When using a thermometer, keep your eye at the same level as the colored liquid to read the degrees.

★ When using a graduated cylinder, you should be aware that liquids create a curved surface known as a **meniscus**. The liquid actually clings to the sides of the container. The scale on the measuring instrument is actually drawn for the lowest point of the curved liquid. You should therefore, measure the amount of liquid from the *lowest point* of the curve.

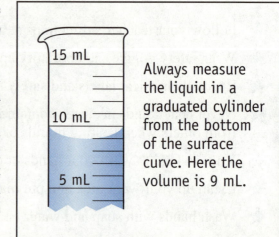

Always measure the liquid in a graduated cylinder from the bottom of the surface curve. Here the volume is 9 mL.

ANALYZE THE DATA

Once a scientist has gathered data from an experiment, it must be analyzed. A scientist looks for patterns in the data. The best way to do this is often to draw a picture from the data. **Graphs** are pictures of data. With a good graph, a researcher can often see a **trend** (*or lack of a trend*) between two variables with one glance.

> ## REPRESENTING DATA
>
> ★ **Bar graph.** A bar graph is a good choice when items are compared.
>
> ★ **Line graph.** A line graph is a good choice when different measurements of the same item are recorded over time.
>
> ★ **Pie chart.** A pie chart is a good choice when pieces of a whole are shown.

The table on the right shows the various heights of two plants in an experiment. Plant A was subjected to two hours more sunlight each day than Plant B. The scientist is investigating the effects of sunlight on each plant.

Now, use the data from the table on the right to show these results in both a *bar graph* and a *line graph*.

★ In the bar graph, show the total height of each plant after the four weeks.

★ In the line graph, show how the plants changed over time.

Week	Plant A	Plant B
Beginning	10 cm	10 cm
Week 1	11 cm	10.5 cm
Week 2	12.2 cm	11 cm
Week 3	13.5 cm	11.5 cm
Week 4	15 cm	12 cm

Title of your bar graph: _____

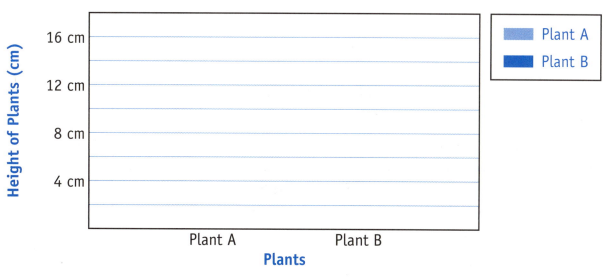

Title of your line graph: _____

TRENDS

Scientists often use the following special terms to describe trends in their data:

★ **Range** — values in data from the lowest to the highest result.

> *A range of temperatures from 0° to 120° C*

★ **Average** — the typical result: to calculate the average, add all the results and divide this amount by the number of results.

> *the average of 10, 13, and 22 cm = 15 cm*

★ **Frequency** — the number of times something is repeated in a period of time.

> *20 waves per second*

★ **Percentage** — a proportion or part of something expressed in hundredths.

> *85% = 85/100*

APPLYING WHAT YOU HAVE LEARNED

✦ A scientist measured the height of 5 bean plants after 4 weeks in an experiment. Here are the heights of the plants as the scientist recorded them:

A	B	C	D	E
11 cm	15 cm	10 cm	18 cm	20 cm

- What is their average height? _____
- What is the range of heights? _____
- What percentage of the plants are 15 cm or taller? _____

Scientists sometimes **extrapolate** data. This means they look at the trends in the data to fill in gaps in their results or to extend their results. For example, suppose a scientist gave one bean plant 5 grams of fertilizer, and gave a second plant 20 grams of fertilizer:

Fertilizer	Growth
5 grams	1 centimeter
20 grams	4 centimeters

How much growth would you expect in this kind of plant from 10 grams of fertilizer? Based on the existing data, a scientist would predict this growth would be more than 1 centimeter but less than 4 centimeters.

In analyzing data, be sure to round off numbers to the *least precise measurement* you record. Remember: your results cannot be more precise than the measurements you have taken.

FORM CONCLUSIONS

Scientists draw **conclusions** based on the results of an experiment and their analysis of the data. These conclusions should relate back to the hypothesis in some way.

What Makes A Good Conclusion?

★ **Logical.** A good conclusion must be logical.

★ **Accurate.** It must correctly report the data.

★ **Supported.** It should be supported by scientific knowledge and by the data from the investigation.

★ **Related to the Hypothesis.** A good conclusion should support or reject the hypothesis, or suggest changes in the hypothesis for further study.

The hypothesis, in turn, is usually based on some larger scientific theory. Almost every scientific investigation therefore helps to support, reject, refine or change some scientific theory. You will learn more about the role of scientific theories in the next chapter.

APPLYING WHAT YOU HAVE LEARNED

✦ Why is it so important that experimental conclusions relate to the hypothesis?

COMMUNICATE RESULTS

Once a scientist completes an experiment, the results must be communicated to others. Usually this takes place in the form of a written report or article.

Sometimes scientists give oral presentations at meetings. A good report or presentation uses clear language, includes accurate data and graphs, and includes the methods and procedures followed so other scientists can *repeat* the experiment and verify the results. This encourages others to build on what was learned.

A good scientist communicates results to other scientists orally or in a written report

Sometimes disagreements develop over an experimenter's conclusions or interpretations of the data. This process of questioning, disagreement and debate increases our scientific knowledge. Good criticism can help refine an investigator's questions or procedures. Open communication is essential for scientific knowledge to advance. Scientists may not always reach the same conclusions, but they generally accept the value of scientific investigation and open debate.

APPLYING WHAT YOU HAVE LEARNED

✦ Why is it so important that experiments can be repeated by other scientists?

INVESTIGATE FURTHER

After drawing conclusions and communicating results, a scientist often thinks of new questions for further research. For example, which *types* of nitrogen-based fertilizers would promote the *fastest* growth in bean plants?

CHAPTER 3: THE PROCESS OF SCIENTIFIC INVESTIGATION   33

WHAT YOU SHOULD KNOW

A. You should know that **scientific inquiry** asks questions about the natural world in a systematic way and guides the growth of scientific knowledge.

B. You should know that a **scientific theory** attempts to explain data received from observing nature and conducting experiments. Models and mathematics help scientists develop theories.

C. You should know how to use the following steps to conduct an experiment:

★ Design an experiment to test your hypothesis. Determine the independent and dependent variables in your experimental design. Hold other variables constant during the experiment.

★ Take safety into account before beginning any experiment. Have safety equipment on hand like safety goggles, gloves, and a fire extinguisher.

★ Measure your results. Use lab instruments to take accurate and precise measurements. The results you record should be no more precise than your least precise measurement.

★ Relate experimental results to your original hypothesis in your conclusion. Your results should lead you to support, reject or change your hypothesis.

★ Communicate your results to others.

CHAPTER STUDY CARDS

Scientific Inquiry

★ **Scientific Inquiry.** Scientists ask questions about what they observe in the natural world. Inquiry guides research.

★ **Theory.** A possible explanation of observations and data, which is tested and revised.

★ **Experiments.** An experiment tests a hypothesis in a controlled setting, usually by changing one variable to see how it affects another variable. Experiments produce much of the data that make up modern science.

Steps in a Scientific Investigation

★ Ask Scientific Questions
★ Form a Testable Hypothesis
★ Design an Experiment to Test the Hypothesis
★ Select and Use Equipment and Technology
★ Collect Data
★ Analyze Data and Form Conclusions
★ Communicate Results
★ Investigate Further

Laboratory and Field Terminology

★ **Models.** A model is a diagram or object that represents something and shows how it works.

★ **Hypothesis.** An educated guess that tries to answer a question under investigation.

★ **Independent Variable.** A variable that is changed to find how that change affects other variables in the experiment.

★ **Dependent Variable.** A variable that is modified as the result of a change to an independent variable in the experiment.

Laboratory and Field Experiments

★ **Laboratory Equipment.** Should provide precise measurements. These tools include a balance, meter stick, thermometer, graduated cylinder.

★ **Laboratory Safety.** Safety should always be an important concern in all experiments.

★ **Experimental Group vs. Control Group.** The "control group" is not subject to changes in the "independent variable." Changes in the experimental group are then contrasted with the control group.

CHECKING YOUR UNDERSTANDING

1 The diagram below shows two germinating corn seeds that have been placed in identical bottles and kept in the dark. Bottle A will be rotated 90° each day for the next 6 days. Bottle B will not be rotated.

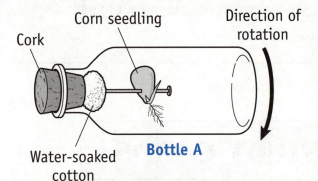
Bottle A

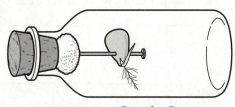
Bottle B

Which hypothesis is tested in this experiment?

A Water is needed for proper plant growth.
B Gravity affects plant growth.
C Enzymes promote seed development.
D The amount of light received affects chlorophyll production.

♦ Examine the Question
♦ Recall What You Know
♦ Apply What You Know

OBJ. 1
8.2 (A)

HINT

This question examines your understanding of the methods of scientific investigation. You should recall that an experiment usually tests the effects of changing one "independent" variable. In this question, only one variable is changed: whether the bottle is turned or not. There is no light in the room, and there is no difference in the water or enzymes. Therefore, the answer is B. The experiment tests the effects of gravity on the growth of the seedlings.

> **Now try answering some additional questions about the process of scientific investigation.**

2 A student is investigating the internal organs of a frog. Which piece of equipment should the student use to move the internal organs aside without damaging them?

   F   A glass beaker
   G   A dissecting probe
   H   A pH meter
   J   A meter stick

   OBJ. 1
   8.4 (A)

3 A scientist is conducting an experiment on plant growth. To find out how tall a plant grew each day, the scientist would —

   A   look at a leaf from the plant under the microscope
   B   put the plant in a sunny place for an hour each day
   C   measure the plant with a meter stick at the same time each day
   D   put the plant on a spring scale each day and weigh it

   OBJ. 1
   8.2 (A)

4 A biologist reports a new discovery based on experimental laboratory results. If the experimental results are accurate, biologists in other laboratories should be able to repeat the same experiment —

   F   using different variables and get similar results
   G   and get different results
   H   and get similar results
   J   under different experimental conditions and get similar results

   OBJ. 1
   8.2 (D)

5 An experimental design included references to prior experiments, a list of materials and equipment, and detailed step-by-step procedures. What else should be included before the experiment can be started?

   A   A set of data
   B   A conclusion based on the data
   C   An inference based on results
   D   A list of appropriate safety precautions

   OBJ. 1
   8.1 (A)

6 A student is conducting an experiment to find out what factors affect plant growth. The student could change the amount of light, heat, water and type of the soil. However, during the investigation, the amounts of light, heat and water remain the same. Only the type of the soil is changed. What is the independent variable in this experiment?

   F   Plant growth
   G   Water
   H   Light
   J   Soil type

   ♦ Examine the Question
   ♦ Recall What You Know
   ♦ Apply What You Know

   OBJ. 1
   8.2 (A)

7 When heating a solution in a test tube, a student should —

    A point the test tube in any direction
    B hold the test tube with two fingers
    C cork the test tube
    D wear safety goggles

OBJ. 1
8.1 (A)

8 Which sentence states a testable hypothesis?

    F Environmental conditions affect plants more than animals.
    G Boil 100 milliliters of water, let it cool, and then add 10 seeds to the water.
    H The temperature of the water in a lake is related to its depth.
    J A lamp, two beakers, and elodea plants are selected for the investigation.

OBJ. 1
8.2 (A)

9 A new drug for the treatment of asthma is tested on 100 people. The people are evenly divided into two groups. One group is given the drug, and the other group is given a sugar pill. In this experiment, the group that is given the sugar pill serves as the —

    A experimental group
    B control group
    C independent variable
    D dependent variable

♦ Examine the Question
♦ Recall What You Know
♦ Apply What You Know

OBJ. 1
8.2 (A)

10 An experiment was performed to determine the effect of different minerals on plant growth. Forty pots containing identical plants were divided into four equal groups and placed in a well-lighted greenhouse. Each pot contained an equal amount of non-mineral soil and one plant. Minerals were then added in equal amounts to each experimental group of pots as shown below:

Control Group	Experimental Groups		
	Water + Nitrogen salts	Water + Potassium salts	Water + Phosphorus salts

What should be added to the control group of pots?

    F Water   H Nitrogen salts
    G Potassium salts   J Potassium and phosphorus salts

OBJ. 1
8.2 (A)

11 A graduated cylinder is filled with water to 28 mL. A rock is placed in the cylinder, and the water rises to the 57mL mark. What is the volume of the rock?

    A 57 mL
    B 85 mL
    C 29 mL
    D cannot be determined

♦ Examine the Question
♦ Recall What You Know
♦ Apply What You Know

OBJ. 1
8.2 (C)

12 The piece of laboratory equipment below can be used to measure the —

    F temperature of a flame
    G volume of a liquid
    H mass of a powder
    J pressure applied by a gas

OBJ. 1
8.4 (A)

13 Which of the following pieces of laboratory equipment can be safely used to clamp a test tube before heating it?

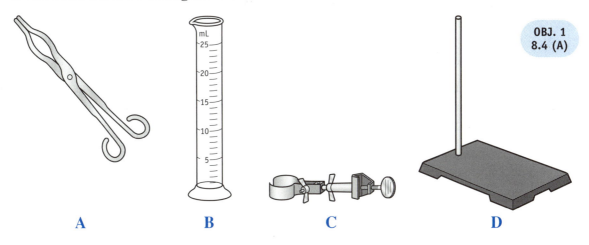

    A   B   C   D

OBJ. 1
8.4 (A)

14 The following balance measures mass in grams. What mass is shown with correct precision?

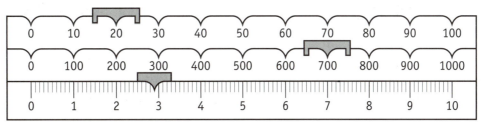

    F 722.90 g
    G 722.9 g
    G 722 g
    J 723 g

OBJ. 1
8.2 (B)

CHAPTER 4

THE GROWTH OF SCIENTIFIC UNDERSTANDING

The last chapter focused on methods of scientific investigation. This chapter looks at how scientific investigation contributes to the growth of scientific understanding.

Scientists first observe the natural world and ask questions. They often look for general rules or principles to explain events. They may build models to represent a natural process or object. Scientists then conduct field and laboratory investigations to test their theories. They share the results of their investigations to increase our understanding of the natural world.

MAJOR IDEAS

A. Scientists use **models** to represent the natural world. By examining how a model works, scientists often can develop **theories** to explain what they observe in nature.

B. Scientists use the results of their field and laboratory investigations as evidence to confirm or revise their theories. In this way, our scientific understanding of the world gradually improves.

C. Scientists are often able to use scientific knowledge to develop new and better products and services.

THE ROLE OF MODELS IN SCIENCE

You learned in earlier chapters that scientists sometimes use models as a way of understanding what happens in nature. They may use a model if the object is too small to be seen, like an atom, or if the object is impossible to look at, like the center of Earth.

CONSTRUCTION OF A MODEL

Modeling is a way for scientists to put complex scientific data into understandable terms. Scientists sometimes construct their models to be three-dimensional. Models often use simple objects to help viewers understand complex principles and relationships. For example, a scientist might construct a model that represents Earth, moon, and the Sun as small spheres that can be moved. This model can illustrate the different phases of the moon or eclipses.

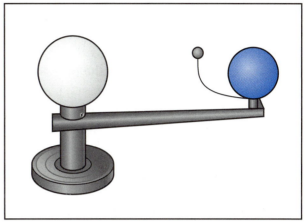

A model of the sun, moon and Earth.

A model can also be a diagram or even just a mental image. For example, scientists think of the atom as a nucleus surrounded by orbiting electrons. Some models of atoms show the electrons as circulating in electron clouds.

THE PURPOSE OF A MODEL

The purpose of a model is to help scientists see relationships and test ideas. As scientists learn more details about what they are investigating, they often can improve the model. For example, they may change the sizes of different parts of the model or the distances between these parts. The improved model helps scientists better picture what is happening. From the model, they can also make predictions about what will happen. Then they can run tests to see if those predictions come true. The more closely a model resembles what it represents, the better its predictions will generally be.

APPLYING WHAT YOU HAVE LEARNED

✦ Think of an object or process that you have learned about in science class this year.
  - How could you make a model of this object or process?
  - Does this model make the object or process easier to understand? If so, explain how.

SCIENTIFIC KNOWLEDGE AND THEORY

You already know that a "big idea" in science is called a theory. A **theory** attempts to explain all of the known information about something from both experimentation and the direct observation of nature.

The Aristotelian View of the Universe. Scientific theories affect how we understand the natural world. For example, Aristotle once taught that Earth was the center of the universe. This theory, however, could not explain all the observations that people made of the skies at night. At times, the planets seemed to move backwards. People made more complicated models, showing the planets moving around in smaller circles within their orbits.

Copernican Theory of the Universe. In the 1500s, Nicolas Copernicus came up with a better idea. He theorized that Earth moved around the Sun. This explained why planets observed from Earth sometimes seemed to move backwards. This new theory provided a more accurate model of our solar system. People at first resisted Copernicus' idea. Eventually, as more data was collected, they began to see that it was correct. Today, from the observations of telescopes and photographs taken from outer space, we know that Earth circles the Sun. This theory is generally accepted even though, when we look at the rising and setting of the Sun, it still might seem as though the Sun is orbiting around the Earth.

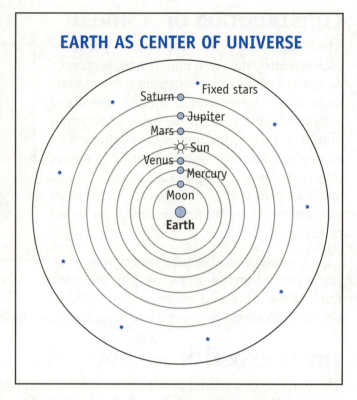

EARTH AS CENTER OF UNIVERSE

Copernicus

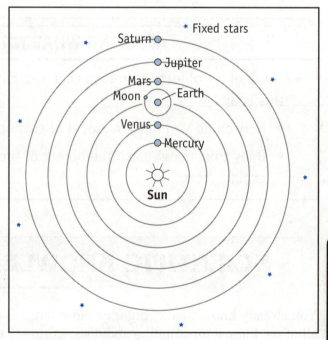

Theories Explain Data. A scientific theory acts as a framework to explain data in a logical way. For example, Sir Isaac Newton studied the movements of planets and objects falling on Earth and saw patterns in the data. Then he came up with the theory of *gravity*. He said that all objects were attracted to each other by a universal force. Newton theorized that the strength of this force was related to the size of the objects and the distance between them. At first, no one could be sure if Newton was correct. Scientists had to check and see if Newton's theory could explain existing observations of planetary movements.

Scientists also tested Newton's ideas by measuring the movements of falling objects. People soon concluded that his theory provided a good explanation.

Theories Guide Research. Theories also guide research. For example, the French scientist Louis Pasteur found bacteria in spoiled food. He theorized that these bacteria caused the food to spoil. From this theory, he predicted that if he applied heat to food, it would kill the bacteria and keep the food from spoiling. Pasteur found that by heating milk for several minutes, harmful microorganisms were killed. His predictions turned out to be correct.

One way to test a theory is to perform experiments and see if the theory can explain the results. Theories must survive such repeated testing. Otherwise, the theory must be rejected or changed.

Louis Pasteur

Analyzing the Strengths and Weaknesses of a Theory. A scientific theory attempts to explain how the natural world works. When you learn about a theory, you must decide how well it is supported. First, consider if the theory seems logical. Does it make common sense? Then see how well it is supported by scientific evidence. If the results of repeated experiments support the theory, then the theory may provide a good explanation of nature. It is able to explain all the known data. For example, scientists now believe that Pasteur was correct in thinking that microscopic bacteria cause food to spoil.

A theory has a weakness if there is data it cannot explain. If some results are different from what the theory predicts, then the theory is not completely accurate and may have to be revised. Therefore, when you learn about a new theory, you should conduct library and Internet research to learn what other scientists think about it. See if the theory is logical and well supported by existing scientific evidence. Based on this support, you should be able to identify the strengths and weaknesses of the theory — what data it can explain, and what it cannot explain.

APPLYING WHAT YOU HAVE LEARNED

♦ Think of a theory you know about and explain how well it is supported.

HOW SCIENTIFIC KNOWLEDGE GROWS

There are many reasons why scientific explanations change and advance. Scientific instruments are regularly improved, and new instruments are sometimes invented. These instruments enable scientists to discover new facts about the world that increase our scientific knowledge. For example, scientists could not see microscopic organisms until the invention of the microscope, and they could not see the moons of other planets until the invention of the telescope. The more that scientists learn, the more they refine their explanations of what we see in the world, and the better we understand how nature really works.

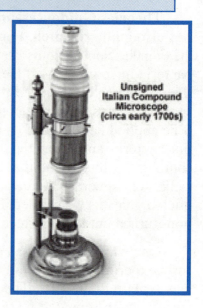
Unsigned Italian Compound Microscope (circa early 1700s)

Scientists continually make new observations about the world and ask questions based on their observations. As new data is collected, scientists often come up with new ideas. They develop better models and theories to explain nature.

APPLYING WHAT YOU HAVE LEARNED

♦ Identify an example in which improved observations helped lead to new scientific ideas.

SCIENCE AND TECHNOLOGY

As you can see, **science** is based on our desire to understand the world around us. Closely related to science is technology. **Technology** is the application of scientific knowledge to develop tools, materials and processes to help us meet our needs. The growth of scientific knowledge often leads to the development of new technologies that allow people to create better products and services.

Many of the modern technological wonders we now use in everyday life have resulted from scientific research.

★ **Television.** Scientists discovered that the ability of selenium to conduct electricity changed with the amount of light. This discovery led to the invention of the television set.

★ **Microwaves.** A scientist working with an electron tube noticed that a candy bar in his pocket melted. He used this discovery to build the first microwave oven.

Robotic arms, developed through scientific research, are now able to perform surgery.

EVALUATING NEW PRODUCTS BASED ON SCIENTIFIC RESEARCH

Scientific advances occur faster today than at any other time in history. These advances provide the foundation for rapid technological change. Many new products now make claims based on scientific research. Such claims, while often true, must be viewed carefully. When you come across a new product or service that claims to be based on new scientific knowledge, use your critical thinking skills to determine if those claims are accurate. Ask yourself the following questions about any claims or data that you read about a new product:

★ *Were the methods used to obtain the data clearly stated, and are they repeatable?*
If you do not know how the data was obtained, the claims about the product may not be accurate or truthful.

★ *Have different groups of scientists obtained the same data?*

★ *Does the evidence support the claims made about the product or service?*
If the data and observations are true, does this mean that the claims made about the product are also true? You should be sure the connection is clear and logical.

★ *Was the sample used large enough to obtain accurate results?*
The larger the sample, the more representative and accurate it is.

★ *Are there other side effects or negative results not reported?*
Has the maker of the product told you everything you should know?

From the information you read about a product, you should be able to draw conclusions about when the product should be used and how to use it safely. For example, you should not use a product that is highly flammable near a fire or next to a source of great heat.

WHAT YOU SHOULD KNOW

★ You should understand that a scientific model represents processes or objects in the natural world, and that any given model might be improved.

★ You should know how to read and analyze scientific explanations, including theories and hypotheses.

★ You should be able to determine the strengths and weaknesses of a scientific explanation, based on how well it is supported by scientific data and observations.

★ You should be able to make predictions based on data and scientific explanations.

★ You should know how scientific theories are tested and revised.

★ You should be able to draw conclusions from data about products and services.

APPLYING WHAT YOU HAVE LEARNED

◆ Now that you have completed this chapter, how would you define "scientific knowledge"?

CHAPTER STUDY CARDS

Scientific Knowledge

★ Scientific knowledge changes over time as scientists ask new questions, conduct new experiments, and revise their theories.
- **Isaac Newton.** Studied how planets and other objects fall to Earth. From this, Newton developed the law of gravity.
- **Louis Pasteur.** Using a microscope, Pasteur theorized bacteria cause spoilage.

★ **Analyzing Theories and Hypotheses.** When you learn about a scientific theory or hypothesis, see how well it is supported by logic and scientific evidence (observations and data).

Evaluating Products and Services

When looking at data or claims about a product:

★ Look to see how the data was obtained.

★ See if the data supports the claims.

★ Think about side-effects or negative results that are not reported.

★ Think about how the product can be used safely.

CHECKING YOUR UNDERSTANDING

Students placed samples of the same type of bacteria in 10 identical petri dishes with the same nutrients. They placed equal amounts of a chemical in 5 of the dishes and nothing in the other 5 dishes. They checked all 10 dishes under the microscope after five days.

1. **The hypothesis for this experiment was probably that the chemical would —**
   A change the color of the petri dishes
   B affect the growth of the bacteria
   C disappear
   D combine with the nutrients

   ♦ Examine the Question
   ♦ Recall What You Know
   ♦ Apply What You Know

   OBJ. 1
   8.2 (A)

This question examines your ability to analyze information in an experiment. Every experiment usually involves the testing of some hypothesis. In this experiment, a chemical was introduced to five groups of bacteria but not to the other five groups. To answer the question, you must select the hypothesis you think the experiment is testing. The experimental groups have been exposed to the chemical but the control groups have not. Therefore, the experiment tests whether the chemical will affect the growth of the bacteria.

Now try answering some additional questions on your own about scientific theories and knowledge.

TEMPERATURE OF LAKE TRAVERSE AT VARYING DEPTHS

Depth in Meters	Temperature in Celsius
1	40°
2	35°
3	30°

2. **According to the data on the table, the temperature in Lake Traverse at 4 meters is most probably —**
   F 35° C
   G 30° C
   H 25° C
   J 20° C

   OBJ. 1
   8.4 (B)

> Jupiter is the largest planet in the solar system. Saturn is the second largest planet. Some scientists have theorized that when these planets are viewed through a telescope, what we see is mainly their atmospheres — the mixture of gases that surrounds them.

3 Which of the following would provide the best evidence in support of this theory?

    A Jupiter and Saturn are the two largest planets in the solar system.
    B Unmanned satellites sent to these planets find deep atmospheres.
    C Jupiter and Saturn are farther from the Sun than Earth.
    D Jupiter and Saturn each have a large number of moons.

OBJ. 1
8.3 (A)

4 Which conclusion could reasonably be made about Texas Cola?

    F It is not a good source of calcium.
    G It contains harmful bacteria.
    H It is good for your teeth.
    J It provides a healthy breakfast.

OBJ. 1
8.3 (B)

> A teacher takes a class on a field trip to a nearby pond. To show his class how the human ear hears sounds, the teacher drops a stone into the pond. The stone hitting the water creates small waves in the pond. When the waves hit a leaf floating at the edge of the pond, the leaf begins to move back and forth.

5 In this model, the leaf represents —

    A a sound wave
    B compressed air
    C the human brain
    D an ear drum

♦ Examine the Question
♦ Recall What You Know
♦ Apply What You Know

OBJ. 1
8.3 (C)

CHECKLIST OF OBJECTIVES IN THIS UNIT

*At the end of each content unit you will find a **Checklist of Objectives** like the one below. The purpose of these checklists is to help you mentally review the major objectives examined in the unit before moving on to the next unit.*

Directions. Now that you have completed this unit, place a check (✔) next to those objectives you understand. If you are having trouble recalling information about any of these objectives, review the chapter listed in the brackets.

You should be able to:

❑ plan and implement investigative procedures including asking questions, formulating testable hypotheses, and selecting and using equipment and technology. **[Chapter 3]**

❑ demonstrate safe practices during field and laboratory investigations. **[Chapter 3]**

❑ collect, record, and analyze information using tools including beakers, petri dishes, meter sticks, graduated cylinders, weather instruments, hot plates, dissecting equipment, test tubes, safety goggles, spring scales, balances, microscopes, telescopes, thermometers, calculators, field equipment, computers, water test kits, and timing devices. **[Chapter 3]**

❑ collect data by observing and measuring. **[Chapter 3]**

❑ construct graphs, tables, maps, and charts using tools to organize, examine, and evaluate data. **[Chapter 3]**

❑ communicate valid conclusions. **[Chapter 3]**

❑ represent the natural world using models, and identify their limitations. **[Chapters 3 and 4]**

❑ analyze, review, and critique scientific explanations, including hypotheses and theories, as to their strengths and weaknesses, using scientific evidence and information. **[Chapter 4]**

❑ use collected information to make predictions. **[Chapter 4]**

❑ draw inferences based on data for products and services. **[Chapter 4]**

UNIT 3: THE STRUCTURE AND PROPERTIES OF MATTER

In this unit, you will learn what you need to know about the nature of **matter** — all things that take up space and have mass. Just look around you. Anything that you can see and touch is some form of matter.

Did you know that all forms of matter share a similar structure? They are all made of tiny particles called atoms. Even though all forms of matter are made up of atoms, they are not the same. Differences among atoms give various forms of matter their different properties. These atomic differences also determine how some atoms will combine with other atoms to create new substances.

Scientists study matter by investigating the properties of different substances.

★ **Chapter 5: The Structure of Matter**
In this chapter, you will learn how each atom is made up of the same parts — protons, neutrons and electrons. You will also learn how different atoms combine in chemical reactions.

★ **Chapter 6: The Properties of Matter**
In this chapter, you will examine the physical and chemical properties of different forms of matter, and learn how to read the Periodic Table of the Elements.

CHAPTER 5

THE STRUCTURE OF MATTER

In this chapter, you will explore the structure of matter.

MAJOR IDEAS

- A. **Matter** is anything that has mass and takes up space.
- B. All matter is made up of minute particles known as **atoms**. Atoms are composed of protons, neutrons, and electrons.
- C. An **element** is a substance made up of only one type of atom.
- D. Atoms can react with other atoms to form compounds. A **compound** is made up of two or more different types of atoms.
- E. Both *before* and *after* a chemical reaction, the number of atoms and the total mass remain the same.

Matter is the stuff of the universe. Anything that has **mass** and takes up space is known as matter. Matter comes in many different shapes and sizes. For example, air and water are matter. This book is matter. However, not everything is matter. Light and electricity are not matter because they do not have mass and do not take up their own separate space.

APPLYING WHAT YOU HAVE LEARNED

✦ Create a chart that identifies different examples of matter and things that are not matter. Label the first column "Matter," and the second "Not Matter." Provide three examples of each.

Matter	Not Matter
1. _____	1. _____
2. _____	2. _____
3. _____	3. _____

> ### MASS OR WEIGHT?
> **Mass** is the *amount* of matter an object has. Scientists usually measure mass in **grams** (g) or **kilograms** (kg). On Earth, the mass of an object gives it weight. **Weight** is the force of attraction created by gravity. Our weight changes if we go to the moon or some other planet. On the moon, for example, you would weigh less. However, your mass on the moon is the same as it is on Earth.

ATOMS AND THEIR SUBATOMIC PARTS

All matter is made up of minute particles called **atoms**. Scientists see these atoms as the basic building blocks of matter. Scientists once thought that atoms were the smallest parts of matter. In the 20th century, they discovered that atoms could be broken up into even smaller **subatomic particles**. The three most important of these particles, found in almost all atoms, are:

PROTONS

The **proton** is a particle with a **positive** (+) electrical charge and a mass of **1 atomic mass unit**. Protons are located in the **nucleus** of the atom. The **nucleus** is the dense center of the atom.

NEUTRONS

The **neutron** is a neutral particle with **no** electrical charge. It has approximately the same mass as the proton (*1 atomic mass unit*). Neutrons are also located in the nucleus.

ELECTRONS

Electrons are small particles with a **negative** (−) electrical charge. They move around the nucleus at very high speeds. Compared to protons and neutrons, they have no significant mass. No one can actually measure the speed, direction, and location of an electron. However, scientists believe electrons move in different **energy levels** or **shells** around the nucleus.

Atoms are so small that their parts cannot be seen, even when a powerful microscope is used. However, scientists use diagrams or models to explain how they behave: protons and neutrons are joined together in the **nucleus**, or the central part of the atom. Electrons are shown orbiting the nucleus in their shells, or more accurately, as **electron clouds**. The electron moves around the cloud, almost randomly, at very high speeds.

CHAPTER 5: THE STRUCTURE OF MATTER   51

An older model of the atom shows electrons orbiting the nucleus like planets in the solar system.

A more recent model of the atom shows electrons moving around the nucleus in electron clouds.

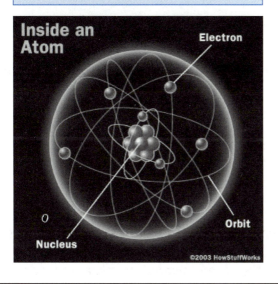

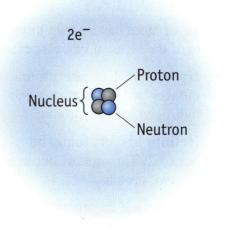

APPLYING WHAT YOU HAVE LEARNED

✦ At a ball game, people sometimes use a scorecard to keep track of what each player has done during the game. Complete the following scorecard to help you recall the differences between an *electron*, *proton* and *neutron*:

	Electron	Proton	Neutron
• Electrical Charge:			
• Mass:			
• Location:			

The diagrams below help us to understand how atoms act. In reality, the electrons are actually much smaller and farther from the nucleus than shown on the diagram. If the nucleus were the size of a marble, the electrons would be a whole football field away.

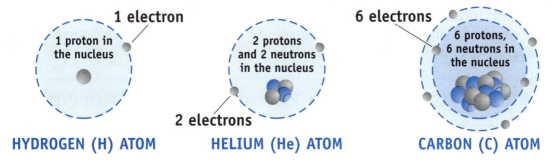

HYDROGEN (H) ATOM   HELIUM (He) ATOM   CARBON (C) ATOM

ATOMIC SYMBOLS

To identify each type of atom, or **element**, scientists use a symbol of one or two letters, based on its name. The first letter is always capitalized, but not the second. For example, the symbol for hydrogen is **H**, and the symbol for carbon is **C**. The symbol for helium is **He**, while the symbol for iron is **Fe**.

ATOMIC NUMBERS

Every type of atom has its own unique **atomic number**. This equals the number of protons each atom of the element has. For example:

★ **Hydrogen** has an atomic number of 1. This means that hydrogen has one proton.

★ **Helium** has an atomic number of 2. This means helium has two protons.

★ **Carbon** has an atomic number of 6 because it has six protons.

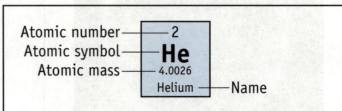

Every type of atom is shown on the Periodic Table of the Elements. For each element, the table displays the atomic symbol, number, and mass.

ELECTRICAL CHARGE

When an atom is not combined with other atoms, it is electrically neutral. This means it has no electrical charge. This is because an uncombined atom always has the same number of positively charged protons and negatively charged electrons. As a result, the atomic number also tells the number of electrons. These are always the same as the number of protons.

> **ATOMIC NUMBER = NUMBER OF PROTONS = NUMBER OF ELECTRONS**

When combined with other atoms, atoms sometimes give away or take electrons from neighboring atoms. When an atom gives away or gains electrons, it takes on a positive or negative electrical charge.

ATOMIC MASS

Scientists measure the mass of an atom by its **atomic mass units**. The two particles with mass in an atom are its protons and neutrons. An electron has almost no significant mass. To determine the **mass** of a single atom — the amount of matter it contains — scientists simply add the number of its protons and neutrons together.

> **ATOMIC MASS = NUMBER OF PROTONS + NUMBER OF NEUTRONS**

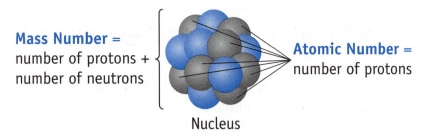

Nucleus

The mass of a typical hydrogen atom is therefore 1; a typical helium atom is 4; and a typical carbon atom is 12. Look back at the diagram on page 51 to see why this is so. A hydrogen atom has only one proton and no neutrons. A helium atom has two protons and two neutrons. A carbon atom has six protons and six neutrons.

If you know the atomic mass and the atomic number of an atom, you can find the number of its protons, neutrons, and electrons. For example, the atomic number of silicon is 14 — the number of its protons. Its atomic mass is 28. Can you guess how many neutrons it has?

ATOMIC MASS – ATOMIC NUMBER = NUMBER OF NEUTRONS

APPLYING WHAT YOU HAVE LEARNED

- An atom of nitrogen has 7 protons, 7 electrons, and 7 neutrons. What is its atomic number and atomic mass? Draw a diagram of it.

- An atom of nickel has an atomic number of 28. It has an atomic mass of 59. How many protons, neutrons, and electrons does it have?

Why aren't the atomic masses shown on the Periodic Table as whole numbers? In nature, elements sometimes have extra neutrons. The atomic mass on the Periodic Table shows the average mass of these different forms of the same element.

WHAT YOU SHOULD KNOW

A. You should know that all matter is made up of minute particles known as **atoms**. Atoms are made up of subatomic particles. **Protons** are found in the nucleus of the atom and have a positive charge. **Neutrons**, also found in the nucleus, are neutral. **Electrons** are negatively charged particles that move around the nucleus at high speeds.

B. You should know that uncombined atoms have the same number of positively charged protons and negatively charged electrons.

C. You should know that scientists use a symbol with one or two letters to identify each type of atom. Its **atomic number** is the number of its protons. Its **atomic mass** is the number of its protons and neutrons.

ELEMENTS AND COMPOUNDS

Elements. An **element** is any form of matter made up of **one** type of atom. An element cannot be divided into any simpler form of matter by chemical means. An atom is the smallest part of an element that has the properties of that element. For example, in an iron rod, every atom in the rod has the same structure. All the known elements are listed in the **Periodic Table of the Elements**. Scientists have found 93 elements in nature, and more than 20 others have been produced in a laboratory.

Compounds. Sometimes different atoms combine together to make compounds. A **compound** is a substance made up of atoms of two or more different elements, chemically joined together. Because these atoms are chemically joined, they always combine in the same proportions. In fact, the atoms in a compound are either sharing or exchanging some of their electrons. This enables their atoms to fill up their outer energy levels of electrons.

When elements combine to form compounds, the atoms of each element lose their individual properties and take on new properties. These new properties are different from those of the elements that combined to make the compound. For example, water (H_2O) is a compound made up of atoms of hydrogen and oxygen. Its properties are different from those of either oxygen or hydrogen. Other common compounds include sugar ($C_{12}H_{22}O_{11}$), salt (NaCl), ammonia, and carbon dioxide (CO_2).

Molecules. The smallest part of any compound is a molecule. A molecule is a group of several atoms bonded together. The different atoms in a molecule actually share some of their electrons. In a molecule of water, an oxygen atom and two hydrogen atoms share some electrons. This creates a bond that holds the molecule together. The atoms of elements also often combine into molecules (O_2 or H_2).

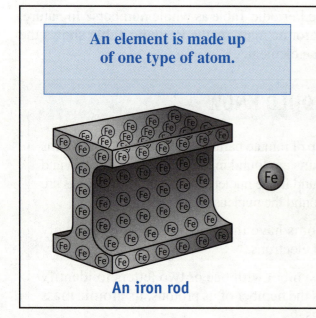

An element is made up of one type of atom.

An iron rod

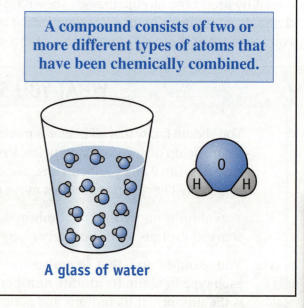

A compound consists of two or more different types of atoms that have been chemically combined.

A glass of water

APPLYING WHAT YOU HAVE LEARNED

✦ Complete the information in the table comparing elements and compounds:

	Elements	Compounds
• How many kinds of atoms does it have?	_____	_____
• Can it be broken down into a simpler form of matter?	_____	_____
• Does it have the same properties as the atoms that make it up?	_____	_____
• Name two examples of each.	1. _____ 2. _____	1. _____ 2. _____

FORMULAS AND CHEMICAL REACTIONS

To write a **chemical formula**, scientists use the atomic symbol to represent each element.

They make use of the same symbols to represent compounds. The atomic symbols of the elements in the compound are written together, like the letters of a word. For example, **NaCl** represents sodium chloride. This compound has both sodium and chlorine atoms. You may know this chemical as table salt.

A small number is written below the line if there is more than one atom of a certain type in the molecule. Each molecule of water, for instance, has two hydrogen atoms and one oxygen atom. This is represented by H_2O.

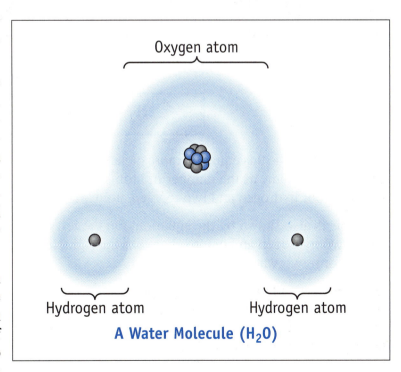

A Water Molecule (H_2O)

UNDERSTANDING CHEMICAL REACTIONS

In a **chemical reaction**, one or more **substances** (*elements or compounds*) change to form new substances. Oxygen and hydrogen atoms, for example, may combine to form water. The substances that *go into* the reaction and those that *come out* of the reaction are different. They have different arrangements of atoms with different properties.

The *number* and *type* of atoms, however, remain the same both *before* and *after* the reaction. Because the number of atoms before and after the reaction is the same, their mass also remains the same. This principle is sometimes known as the **conservation of mass**. For example, when two hydrogen molecules (H_2) are heated with one oxygen molecule (O_2), a chemical reaction occurs. This reaction can be shown as follows:

STEP 1: BEFORE THE REACTION
Oxygen and hydrogen gases are present in separate molecules (H_2 and O_2).

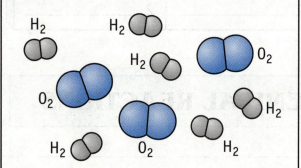

STEP 2: AFTER THE REACTION
The oxygen and hydrogen atoms have combined to form water molecules (H_2O).

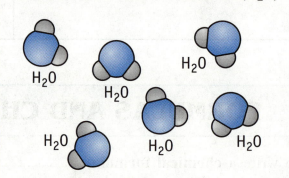

This reaction can be represented by the chemical equation below:

$$2H_2 + O_2 \rightarrow 2H_2O$$

Let's examine this equation more closely. You can see that the number of atoms before the reaction is the same as the number of atoms after it. There are six atoms *before* and *after* the reaction: four hydrogen and two oxygen atoms. However, they are combined differently.

This is also shown by the equation. On the left of the equation, there are three gas molecules: two hydrogen and one oxygen. On the right side, these atoms have combined in a new way to form two water molecules.

The gases on the right side are both **elements**: their molecules are made up of the same kinds of atoms. Water is a **compound**: it has different kinds of atoms combined together. Although water has both hydrogen and oxygen atoms, it is a totally different substance with its own chemical and physical properties.

CHAPTER 5: THE STRUCTURE OF MATTER

READING CHEMICAL EQUATIONS

$$2H_2 + O_2 \rightarrow 2H_2O$$

Scientists write chemical formulas and equations to show what happens in a chemical reaction. When writing equations, they follow certain rules:

★ The substances that go into the reaction are usually written on the left side of the reaction. The arrow indicates the direction of the reaction. What is produced by the reaction comes after the arrow.

★ To indicate the number of **atoms** in each molecule, scientists put a little number below the line after the atomic symbol. In H_2, the small number $_2$ indicates that the hydrogen molecule has two atoms. If there is no number after the atomic symbol, then the number of atoms is one. In H_2O, there is only one oxygen atom.

★ To indicate the number of **molecules** involved in the reaction, scientists write the number in front of the molecule. When there is no number in front of a molecule, the number of molecules is one. In $2H_2 + O_2 \rightarrow 2H_2O$, there are two hydrogen molecules and one oxygen molecule on the left side of the equation.

★ There are always the same number of atoms on both sides of a chemical equation. You can multiply the number of molecules by the number of atoms of each type to check. Here, there are four hydrogen and two oxygen atoms on each side of the equation.

APPLYING WHAT YOU HAVE LEARNED

Now you can try to write a chemical equation. Suppose that a sample of the metal magnesium (Mg) combines with the gas chlorine (Cl) to create the compound magnesium chloride. Each magnesium atom will form bonds with two chlorine atoms. Fill in the blank lines below to write out your chemical equation:

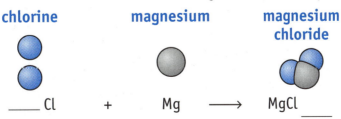

____ Cl   +   Mg $\longrightarrow$ MgCl____

✦ How many chlorine atoms are on each side of your equation? _____

✦ How many magnesium atoms are on each side of your equation? _____

WHAT YOU SHOULD KNOW

A. You should know that an **element** has one type of atom; a **compound** has two or more types of atoms. A **molecule** is the smallest part of a compound with the properties of that compound.

B. You should know that in a **chemical reaction**, one or more substances break apart or combine together to form new substances. The number and type of atoms and the total mass remain the same *before* and *after* the reaction.

CHAPTER STUDY CARDS

Atoms and Subatomic Particles

Atom. All matter is made up of unique particles called atoms. An atom contains:

★ **Protons.** Positively charged; have atomic mass of one; located in the nucleus of the atom.

★ **Neutrons.** Neutral in charge; the same mass as the proton; also located in the nucleus of the atom.

★ **Electrons.** Negatively charged; they move around the nucleus at high speeds; same number as protons.

Atomic Notation

★ **Atomic Symbol.** Scientists use a symbol of one or two letters to identify elements. The first letter is always capitalized, but not the second. Examples: H, O, He, Fe

★ **Atomic Number.** The number of protons an atom has; also equals the number of electrons of the uncharged atom.

★ **Atomic Mass.** The number of protons and neutrons in an atom.

Elements and Compounds

★ **Elements.** Any form of matter made up of identical atoms. Examples: carbon, oxygen.

★ **Compounds.** A substance made up of atoms of different elements joined chemically together. These elements are always combined in fixed proportions. Examples: water, salt, sugar.

★ **Molecule.** A group of atoms that share one or more pairs of electrons; the smallest part of a compound with its properties.

Chemical Reactions

★ **Chemical Reactions.** These are combinations or divisions of substances that result in one or more new substances with new physical and chemical properties.

★ **Chemical Equations.** There should be the same number of atoms on each side of the equation. Multiply the atoms in each molecule by the number of molecules to see that they balance: $2H_2 + O_2 \rightarrow 2H_2O$

★ **Conservation of Mass.** Mass cannot be created or destroyed in a chemical reaction.

CHECKING YOUR UNDERSTANDING

1 **Which correctly describes protons and neutrons?**
  A The same masses and the same electrical charges
  B The same masses and different electrical charges
  C Different masses and the same electrical charges
  D Different masses and different electrical charges

OBJ. 3
8.8 (A)

HINT
This question tests your understanding of subatomic particles. Both protons and neutrons are located in the nucleus. Each proton has an atomic mass of 1 and a *positive* electrical charge. Each neutron has an atomic mass unit of 1 and no electrical charge. Only choice **B** is correct.

Now try answering some additional questions about the structure of matter.

Atom	Number of Protons	Number of Electrons
Sulfur	16	18
Chromium	24	24
Cobalt	27	24
Americium	95	92

2 **According to the chart, which of these atoms is electrically neutral?**
  F Sulfur   H Cobalt
  G Chromium   J Americium

OBJ. 3
8.8 (B)

3 **Which statement is true about the charges assigned to an electron and a proton?**
  A Both an electron and a proton are positively charged.
  B An clcctron is positive and a proton is negative.
  C An electron is negative and a proton is positive.
  D Both an electron and a proton are negatively charged.

OBJ. 3
8.8 (A)

4 **Compared to the entire atom, the nucleus of an atom is —**
  F smaller and has some of the atom's protons
  G smaller and has all of the atom's protons and neutrons
  H larger and has most of the atom's mass
  J larger and has few of the atom's protons and electrons

OBJ. 3
8.8 (A)

The following diagram illustrates what happened in a chemical reaction:

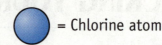
= Hydrogen molecule   = Chlorine atom

BEFORE THE REACTION

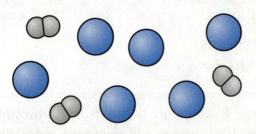

AFTER THE REACTION

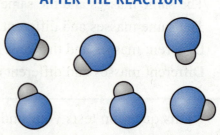

5 Which of the following chemical equations best represents the reaction above?

A $H_2 + 2\,Cl \rightarrow 2HCl$
B $H_2O + 2\,Cl \rightarrow 2HCl + O_2$
C $H_2 + Cl_2 \rightarrow 2HCl$
D $H + Cl \rightarrow HCl$

OBJ. 3
8.9 (C)

6 Which of the following best describes electrons?

F Positive subatomic particles that move around the nucleus
G Negative subatomic particles that move around the nucleus
H Positive subatomic particles located in the nucleus
J Negative subatomic particles located in the nucleus

OBJ. 3
8.8 (A)

7 An element can be described as a substance with —

A different kinds of atoms
B different properties than its individual atoms
C positively and negatively charged atoms
D one kind of atom

OBJ. 3
7.7 (C)

8 An atom of sodium (Na) contains 12 neutrons and 11 electrons. How many protons does this atom have?

F 1
G 11
H 12
J 23

OBJ. 3
8.8 (A)

9 Sugar is a compound. Its chemical formula is $C_{12}H_{22}O_{11}$. What elements are present in sugar?

A Carbon, helium and osmium
B Carbon, hydrogen and oxygen
C Carbon and water
D Carbon dioxide and hydrogen

♦ Examine the Question
♦ Recall What You Know
♦ Apply What You Know

OBJ. 3
7.7 (D)

10 How does a compound differ from an element?

    F It has atoms of more than one type.
    G It has the same properties as its individual atoms.
    H Its atoms do not share electrons.
    J It has a negative electrical charge.

OBJ. 3
7.7 (C)

11 An atom of the element phosphorus (P) has an atomic number of 15 and an atomic mass of 31. How many neutrons does it have?

    A 15   **C** 31
    B 16   **D** none

OBJ. 3
8.8 (B)

12 An atom of oxygen (O) has 8 protons, 8 electrons, and 8 neutrons. What is its atomic mass?

    F 0
    G 8
    H 16
    J 24

OBJ. 3
8.8 (B)

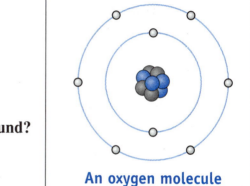
An oxygen molecule

13 Which chemical formula represents a compound?

    A O_2 **C** H_2O
    B Fe   **D** O_3

OBJ. 3
8.9 (C)

14 A molecule has one carbon atom and four hydrogen atoms. Which chemical formula correctly represents this molecule?

    F 4CH
    G CH_4
    H C4H
    J C_3H

♦ Examine the Question
♦ Recall What You Know
♦ Apply What You Know

OBJ. 3
8.9 (C)

15 Which of the following best represents the structure of an atom?

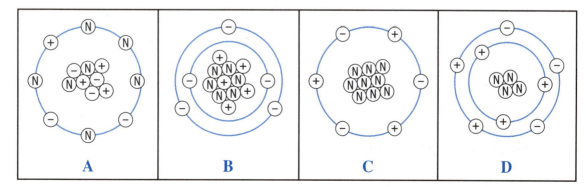

CHAPTER 6

THE PROPERTIES OF MATTER

In this chapter, you will learn about the **properties** (*characteristics*) of different forms of matter. You will see how matter in different forms appears and reacts.

MAJOR IDEAS

A. The **physical properties** of a substance include its color, odor, and hardness. The **chemical properties** of a substance refer to its ability to react with other substances.

B. Matter can exist as a **solid**, **liquid**, or **gas**.

C. **Specific heat** is the amount of heat energy needed to raise the temperature of a substance.

D. The **density** of a substance is found by dividing its mass by its volume.

E. The **Periodic Table of the Elements** lists elements in the order of their atomic number — their number of protons. Elements in the same group often have similar physical and chemical properties.

F. A **mixture** contains two or more substances that are not chemically combined.

PHYSICAL AND CHEMICAL PROPERTIES

Every **substance** (*any element or compound*) has its own set of characteristics or properties that help scientists to set it apart from other substances. These properties fall into two categories — *physical* and *chemical*.

★ **Physical Properties.** When you pick up an object, you often notice certain things about it, such as how much it weighs, how hard it is, and what it looks like. These are examples of physical properties. Every type of matter has its own physical properties. These include its color, odor, density (mass/volume), hardness, melting and boiling point, and how well it conducts heat and electricity.

A physical property can change without changing a substance's chemical make-up —its arrangement of atoms. For example, when ice reaches its melting point and goes from a solid to a liquid, its appearance changes but its chemical composition stays the same.

★ **Chemical Properties.** Chemical properties refer to the ability of a substance to react with other forms of matter. For example, some substances are flammable. If they are heated with oxygen, they will react and burst into flames. The ability of a substance to combine with oxygen is an example of a chemical property.

A student carefully mixes substances.

APPLYING WHAT YOU HAVE LEARNED

✦ How do the physical properties of a substance differ from its chemical properties?

✦ Make a list of two physical properties and two chemical properties of water (H_2O).

Physical Properties	Chemical Properties
1. _____	1. _____
2. _____	2. _____

THE THREE STATES OF MATTER

Every form of matter can exist as either a *solid*, *liquid*, or *gas*. Scientists refer to this condition as the matter's **state**. The state of matter is an example of a physical property. When something changes from solid to liquid, its physical properties change but its chemical properties remain the same. It still has the same atoms.

SOLIDS

Scientists believe that all atoms and molecules are in constant motion. In a solid, atoms and molecules are locked into fixed positions. This gives the substance a fixed volume and shape. Its atoms and molecules still move, but they vibrate in place.

LIQUIDS

When heat energy is transferred to a solid, its particles begin to vibrate more rapidly. Eventually, its atoms or molecules vibrate so strongly they change their position and start to move around each other. The solid melts and becomes a liquid. The temperature at which a solid turns into a liquid is known as its **melting point**.

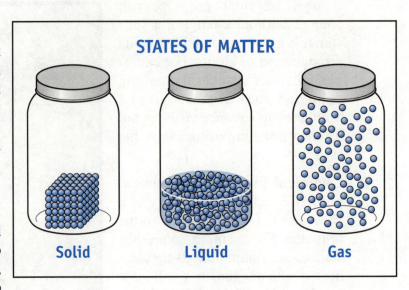

As a liquid, matter may change its shape but still has a fixed volume. Its particles are usually not as tightly packed as in their solid state. Because the particles of a liquid can easily move around each other, the liquid will take the shape of whatever container it is in.

GAS

If heat is applied to a liquid, its atoms and molecules begin to move around more rapidly. Eventually, they break all connections with the other atoms or molecules and spread out in all directions as a **gas**. A gas, therefore, has no fixed shape or volume. The temperature at which a liquid turns into a gas is known as its **boiling point**. The melting and boiling points of a substance are two of its physical properties.

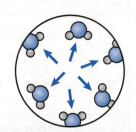

APPLYING WHAT YOU HAVE LEARNED

◆ How do solids, liquids and gases differ from each other? Complete the table by listing their characteristics.

Solids	Liquids	Gases

ENERGY AND CHANGES OF STATE

Temperature is a measure of the speed at which the particles in a substance move. Energy in the form of heat is required to increase the speed of the particles of a substance. Each substance requires a certain amount of heat energy to speed up its particles. The **specific heat** of a substance is the amount of heat needed to raise the temperature of a gram of that substance by one degree Celsius. Heat is measured in **calories**. One gram of water requires one calorie to raise its temperature by 1°C.

Water requires more energy to raise its temperature than most common substances. Its specific heat is high. The same energy it takes to raise a gram of water by 1°C will raise a gram of copper by more than 10°C. Because of this characteristic, bodies of water change temperatures slower than surrounding land areas.

Changes of State
1. Solid to liquid
2. Liquid to solid
3. Liquid to gas
4. Gas to liquid
5. Solid to gas

When a substance changes from solid to liquid or liquid to gas, scientists say that it has undergone a change of state. Changes of state always require or give off energy in the form of heat. For example, if you add heat energy to water, its particles speed up. If you keep adding heat, the liquid water will eventually turn to gas in the form of steam. When extra heat is applied to water that is changing state, its temperature does not increase. Instead, the energy is used to turn the boiling water into steam.

APPLYING WHAT YOU HAVE LEARNED

✦ When snow melts, it turns from a solid to a liquid. Can you give two other examples of changes of state?

1. _____   2. _____

DENSITY

Another physical property of matter is its density or "compactness." The **density** of an object is its mass divided by its volume. This shows how much mass a unit of volume has.

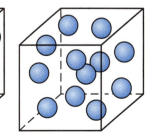

*These two boxes have the same **volume**. Assuming each sphere has the same mass, the box with the greater number of spheres has more mass per unit of volume, or **greater density**.*

A dense substance has more mass in a given space than a less dense substance. For example, lead is more dense than a kitchen sponge. A quantity of lead will have more mass than a sponge of the same volume. Scientists often measure density in grams per cubic centimeter (**g/cm³**).

Ability to Float (Buoyancy). A solid that is less dense than a liquid will **float** in the liquid. For example, a penny will float in mercury because copper is less dense than mercury. The same penny will sink in water because copper has a greater density than water. Similarly, a helium balloon floats in the air because helium is less dense than the atmosphere at Earth's surface.

Water has a density of 1 gram per cubic centimeter. Objects with less density than water will float in water. Objects with greater density will sink in water unless they are given a special shape, like a boat.

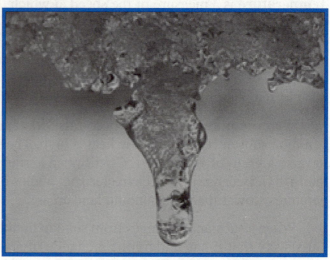

Matter is usually most dense in solid form. Water is an exception. Ice is less dense than liquid water. That explains why ice floats.

APPLYING WHAT YOU HAVE LEARNED

✦ A round object has a density of 1.29 g/cm³ and volume of 2 g/cm³.

- What is its mass? _____
- Will the object float on water? _____

THE PERIODIC TABLE OF THE ELEMENTS

You know that an **element** is any form of matter with only one kind of atom. Each element has its own unique set of physical and chemical properties.

In the mid-1800s, **Dmitri Mendeleev**, a Russian chemist, noticed repeating patterns in the properties of the elements known at that time. Mendeleev developed the **Periodic Table of the Elements** based on those patterns. His table led to the discovery of many new elements.

Dmitri Mendeleev

ORGANIZATION OF THE PERIODIC TABLE

At the bottom of this page you can see the **Periodic Table of the Elements**. In order to understand the table, you need to know how it is arranged:

★ **The Periodic Table of the Elements** lists all the known elements by their symbol. They are arranged in order based on their atomic number — the *number of protons* each atom of that element has. Since hydrogen has just one proton, it appears first in the table.

★ **Periods.** The table is arranged in horizontal rows called **periods**. The period tells you how many electron energy levels the atom has.

★ **Groups.** The vertical columns of the Periodic Table are called **groups**. As you move from left to right, the number of outer electrons in each group increases. The first group has atoms with one outer electron. The last group has atoms with completed outer shells containing eight electrons. The members of each group have similar chemical and physical properties since they contain the same number of outer electrons. For example, the elements in Group 1 combine easily with other elements.

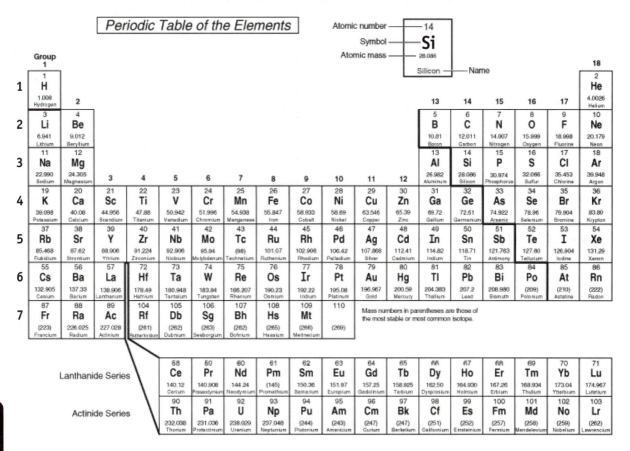

> ### APPLYING WHAT YOU HAVE LEARNED
>
> ◆ In what period do you find iron (Fe)? How many protons does it have?
>
> ◆ In what group do you find magnesium (Mg)? How many electrons does it have?

THE PROPERTIES OF ELEMENTS

The Periodic Table of the Elements helps explain repeating patterns in many of the properties of elements. Based on these patterns, scientists divide the elements into four broad categories.

METALS

Metals make up the largest group of elements. They are found to the left of the bolded line on the table. Metals lose electrons when they combine with other elements.

Metals, except for mercury (Hg), are solid at room temperature. They are generally shiny. Metals are good conductors of electricity and heat. They can be bent or shaped easily, so they can be drawn into wires and hammered into thin sheets. Examples of metals are lithium (Li), sodium (Na), and potassium (K).

Lithium, sodium, and potassium, clockwise from top left

METALLOIDS

The six elements that border the bolded staircase line on the table, such as boron (B) and silicon (Si), are referred to as **metalloids** or **semiconductors**. They have some of the properties of both metals and nonmetals. Metalloids are important because they only partially conduct electricity. They are especially valuable in the semiconductor and computer chip industry. Silicon is a metalloid element used to manufacture computer chips.

NONMETALS

To the right of the bolded staircase are the nonmetals. Examples include carbon (C), nitrogen (N), oxygen (O), phosphorus (P), and sulfur (S). In some ways, nonmetals are the opposite of metals. Their atoms *gain* electrons when they combine with metals to form compounds.

Nonmetals can also form compounds with other nonmetals by sharing electrons. Many nonmetals, such as oxygen and nitrogen, are found in nature as gases. Nonmetal solids like carbon and sulfur are very brittle. They do not bend or twist easily, and do not conduct electricity. Nonmetals are not shiny and do not reflect light.

NOBLE GASES

Group 18, to the far right of the table, consists of the **noble gases** (*helium, neon, argon, krypton, xenon, and radon*). These elements are generally very unreactive or "inert." For example, argon (Ar) is often used to fill light bulbs because it does not easily react. In nature, the noble gases often exist as single atoms.

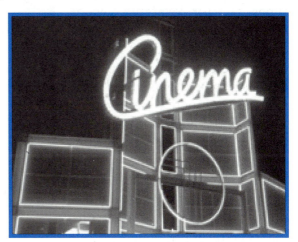

Neon is a noble gas used to light this sign.

SUMMARY: TYPES OF ELEMENTS

	Metals	Metalloids	Nonmetals	Noble Gases
Examples	Sodium (Na), gold (Au), iron (Fe), silver (Ag)	Boron (B), silicon (Si), arsenic (As)	Carbon (C), sulfur (S), oxygen (O), nitrogen (N)	Helium (He), neon (Ne), argon (Ar), xenon (Xe)
Properties	Good conductors of electricity and heat; high melting and boiling points; hard, shiny, and ductile	Share properties of metals and nonmetals; conduct electricity under some conditions	Poor conductors of heat; do not conduct electricity; solids are brittle, non-ductile; many are gases	Not reactive: do not easily combine with other elements
Location on Periodic Table	Left side of the table	Between metals and nonmetals	Right side of the table	Far right of the table (Group 18)

APPLYING WHAT YOU HAVE LEARNED

✦ Compare the properties of metals and nonmetals.

✦ Make a list of five elements in your classroom. Identify each element as a metal, nonmetal, metalloid or noble gas.

MIXTURES AND SOLUTIONS

You already know that **compounds** contain atoms of different types — for example, water (H_2O), or salt (**NaCl**). Although a compound has different types of atoms, their proportions are fixed because these atoms are chemically combined. Compounds always have atoms of each element in the same proportion. Water always has one oxygen atom for every two hydrogen atoms. The elements in a compound cannot be separated without a chemical reaction.

In this experiment, a student mixes various solutions.

A **mixture** contains two or more substances that are mixed together without being chemically combined. The composition of mixtures can be varied.

For example, salt water is a mixture of water and salt. There is no fixed proportion between the ingredients in the mixture. Salt water can have different proportions of water and salt. The substances in a mixture also retain many of their original properties — salt water tastes salty. Finally, the substances in a mixture can be separated without a chemical reaction. They can be separated by physical means. If salt water is boiled, the water will evaporate and the salt will remain.

Element: Sodium (Na)	Compound: Salt (NaCl)	Mixture: Salt water (NaCl, H_2O)
All the atoms are the same.	Different kinds of atoms are uniformly arranged throughout in fixed proportions; has its own unique properties; cannot be separated without a chemical reaction.	Several substances are mixed together without being chemically combined; can be separated by physical means without a chemical reaction.

Solutions. Solutions are **uniform** mixtures. In a solution, small particles of one substance are evenly spread among the particles of the other substance. For example, salt water is a solution. Water molecules surround the atoms of sodium and chlorine in the solution. Only a certain amount of one substance can dissolve in another. If you put too much salt in water, there will not be enough water molecules to surround the sodium and chloride atoms. The excess salt will fall to the bottom without dissolving. Scientists say the solution is then **saturated**.

CHAPTER 6: THE PROPERTIES OF MATTER

WHAT YOU SHOULD KNOW

★ You should know the difference between the **physical** and **chemical properties** of a substance.

★ You should know the differences between the **three states of matter** — *solid, liquid* and *gas*.

★ You should know that the **specific heat** of a substance is the amount of heat needed to raise the temperature of a gram of that substance by 1° Celsius. A substance with a lower specific heat will raise its temperature faster when heated than a substance with a higher specific heat. Water has a higher specific heat than most common substances.

★ You should know how to read the **Periodic Table of the Elements**. The table lists all the elements by their **atomic number** and places them in groups based on similar properties. Metals are on the left side of the table and nonmetals are on the right.

★ You should know that a **mixture** contains different **substances** (*elements and compounds*) that are not chemically combined. Their proportions are not fixed. A **solution** is a uniform mixture in which the different substances are spread evenly throughout the mixture.

CHAPTER STUDY CARDS

Physical & Chemical Properties

Every substance (*element or compound*) has a unique set of properties that allow scientists to tell it apart from other substances.

★ **Physical Properties.** A substance's color, odor, density, melting point and boiling point, ductility, and conductivity.

★ **Chemical Properties.** The ability of a substance to react with other substances.

Three States of Matter

★ **Solids.** Atoms and molecules are locked into fixed positions, giving the substance a fixed volume and shape.

★ **Liquids.** When energy is transferred to a substance, its particles start to move, melting the solid into a liquid. The temperature at which it turns into a liquid is its **melting point**. Liquids have volume but no fixed shape.

★ **Gas.** If more heat is applied to a liquid, its particles will move more rapidly, breaking all connections and turning into a gas. The temperature at which a liquid turns into a gas is its **boiling point**.

Periodic Table of the Elements

Table of elements arranged by their atomic number — the number of protons. An element's position on the table will show many of its general properties

★ **Groups.** Members of the same group have similar chemical and physical properties.
  - Have the same number of outer electrons.
  - Often share other characteristics like the ability to conduct electricity or heat.

Compounds, Mixtures and Solutions

★ **Compound.** Made of two or more elements chemically combined. The elements in a compound are always in the same proportion. The properties of the compound differ from the elements that make it up.

★ **Mixture.** Two or more substances mixed together without chemically combining. The substances can be separated without a chemical reaction.

★ **Solution.** A mixture in which one substance is dissolved uniformly in another substance.

Elements

★ **Metals.** Hard, shiny elements appearing on the left side of the Periodic Table. They are good conductors of heat and electricity. Examples include: iron (Fe), aluminum (Al), and sodium (Na).

★ **Metalloids.** Appear along the bolded line on the Periodic Table. They conduct electricity under some conditions. Examples include: boron (B) and silicon (Si). They are important to the semiconductor industry.

Elements

★ **Nonmetals.** These appear to the right of the Periodic Table. They are poor conductors of heat, and do not conduct electricity. Many are gases or brittle solids. Examples include: carbon (C), oxygen (O), and sulfur (S)

★ **Noble Gases.** Appear to the far right of the Periodic Table (Group 18). They do not usually combine with other substances. Examples include: argon (Ar) and neon (Ne).

CHECKING YOUR UNDERSTANDING

1   The Periodic Table of the Elements can be used by scientists —

   A   to find a list of elements arranged by their atomic mass
   B   to predict how atoms of different elements will combine
   C   to identify all of an element's physical and chemical properties
   D   to determine the differences between compounds

OBJ. 3
8.9 (B)

♦ Examine the Question
♦ Recall What You Know
♦ Apply What You Know

HINT This question examines your understanding of the Periodic Table. Recall that the Periodic Table arranges elements into groups based on their electron arrangement. The Table of Elements can therefore be used to determine how atoms of different elements will combine. This affects their chemical properties.

Now try answering some questions on your own about the properties of matter and the Periodic Table of the Elements.

2   **What is a property of most nonmetals?**

   F   Do not conduct electricity
   G   Conduct electricity well
   H   Shiny and ductile in a solid state
   J   Low specific heat

OBJ. 3
8.9 (B)

3   **The major difference between an element and a compound is that in an element —**

   A   all the atoms are the same type
   B   there are atoms of different types
   C   atoms are combined into molecules
   D   two or more substances are mixed together

♦ Examine the Question
♦ Recall What You Know
♦ Apply What You Know

OBJ. 3
8.9 (C)

4   **Two substances are physically blended together without chemically reacting. They retain their original chemical and physical properties. What is this combination of substances called?**

   F   An element
   G   A molecule
   H   A compound
   J   A mixture

OBJ. 3
8.9 (B)

5   **Which sequence of elements occurs as the atomic numbers increase across Period 3 of the Periodic Table?**

   A   metal ⟶ metalloid ⟶ nonmetal
   B   metal ⟶ nonmetal ⟶ metalloid
   C   nonmetal ⟶ metalloid ⟶ metal
   D   nonmetal ⟶ metal ⟶ metalloid

OBJ. 3
8.9 (B)

6   **One calorie is needed to raise the temperature of one gram of water by 1° Celsius. How many calories are needed to raise the temperature of 10 grams of water from 15°C to 20°C?**

   F   5 calories
   G   10 calories
   H   50 calories
   J   100 calories

OBJ. 3
8.10 (A)

Use the Periodic Table to answer questions 6 – 9 below.

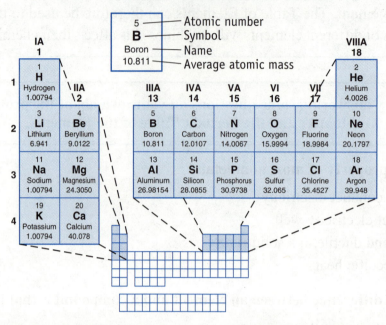

7 Based on the table above, which element probably has physical and chemical properties most similar to boron (B)?

    A Magnesium (Mg)
    B Aluminum (Al)
    C Neon (Ne)
    D Chlorine (Cl)

♦ Examine the Question
♦ Recall What You Know
♦ Apply What You Know

OBJ. 3
8.9 (B)

8 The elements in the Periodic Table are arranged in order of increasing —

    F atomic number
    G hardness in solid form
    H atomic radius
    J neutron number

OBJ. 3
8.9 (B)

9 Which of the following elements is a nonmetal?

    A Fluorine (F)   **C** Calcium (Ca)
    B Magnesium (Mg)   **D** Sodium (Na)

OBJ. 3
8.9 (B)

10 Which two elements are found in CO_2?

    F Carbon and hydrogen
    G Chlorine and hydrogen
    H Carbon and oxygen
    J Chlorine and oxygen

OBJ. 3
8.9 (B)

11 The density of water is 1 g/cm³. Substances whose density is less than that of water will float on the surface of water. Which of the following objects will float on water?

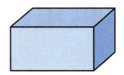

A   a cube with a mass of 45 g, and a volume of 10 cm³

C   a box with a mass of 55 g, and a volume of 100 cm³

OBJ. 3
6.7 (B)

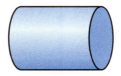

B   a sphere with a mass of 45 g, and a volume of 15 cm³

D   a cylinder with a mass of 120 g, and a volume of 60 cm³

12 **In the Periodic Table of the Elements, the elements of Group 2 can be expected to have —**

   F   similar physical and chemical properties
   G   different chemical properties
   H   the same number of protons
   J   similar densities

OBJ. 3
8.9 (B)

CHECKLIST OF OBJECTIVES IN THIS UNIT

Directions. Now that you have completed this unit, place a check (✔) next to those objectives you understand. If you are having trouble recalling information about any of these objectives, review the chapter in the accompanying brackets. **You should be able to:**

- ❏ describe the structure and parts of an atom. [Chapter 5]
- ❏ identify the properties of an atom, including its mass and electrical charge. [Chapter 5]
- ❏ recognize the importance of formulas and equations to express what happens in a chemical reaction. [Chapter 5]
- ❏ recognize that compounds are composed of elements. [Chapter 5]
- ❏ interpret information on the Periodic Table to understand that properties are used to group elements. [Chapter 5]
- ❏ demonstrate that substances may react chemically to form new substances. [Chapters 5 and 6]
- ❏ classify substances by their physical and chemical properties. [Chapter 6]
- ❏ illustrate interactions between matter and energy, including specific heat. [Chapter 6]

THE NATURE OF MOTION, FORCE AND ENERGY

UNIT 4

In the previous unit, you explored the structure and properties of matter. In this unit, you will learn how matter moves. You will also learn about the forces that move matter.

We constantly observe things in motion. In the 1600s, Sir Isaac Newton developed a series of simple laws that could accurately describe and predict motion. His studies of motion led to a revolution in science.

Electricity is a major form of energy: light bulbs transform it into light; motors convert it into mechanical energy.

Energy provides the force that puts matter into motion. Energy comes in many different forms: heat, electricity, and the energy of moving objects. Like matter, energy cannot be created or destroyed: it only changes from one form to another. In this unit, you will learn about energy as well as motion.

> ★ **Chapter 7: Force and Motion**
> This chapter looks at force and motion. You will learn how motion is measured, and how force can act to change the motion of objects.
>
> ★ **Chapter 8: The Nature of Energy**
> In this chapter, you will learn what energy is, how energy may be either kinetic or potential, and how energy can be transformed from one form to another. You will also learn the properties of energy that moves in waves, such as sound and light waves.

Chapter 7

FORCE AND MOTION

In this chapter, you will learn how matter moves. You will also learn how a force, like the push of a hand, can cause a change in movement.

MAJOR IDEAS

A. **Motion** refers to how an object changes its position over time. Motion can be measured. It consists of both **speed** and **direction**.

B. When an **unbalanced force** is applied to a moving object, it causes that object to change its speed or direction.

C. Simple machines, like levers and pulleys, change the relationship between force and motion.

D. Living things use mechanical force for basic processes, such as the flow of blood through the body.

DESCRIBING MOTION

What exactly is motion? **Motion** refers to the change in position of an object over a period of time. Scientists have developed precise ways of describing and measuring motion and changes in motion.

★ **Distance.** Distance is the total length traveled by a moving object, usually measured in meters (m) or kilometers (km).

★ **Speed.** Speed is the rate of change, or the average distance traveled by a moving object in a given unit of time, such as meters per second (m/s).

$$\text{Speed} = \frac{\text{Distance traveled (m)}}{\text{Time Traveled (s)}} \qquad S = \frac{d}{t}$$

★ **Direction.** Direction describes the location of motion, such as northwards or upwards.

77

For example, Mr. Hidalgo drives his car northwards for two hours. During that time, he travels 100 km in the same direction. His motion can be described as follows:

- **Time:** 2 hours

- **Direction:** north

- **Distance:** 100 kilometers

- **Speed:** 50 km per hour

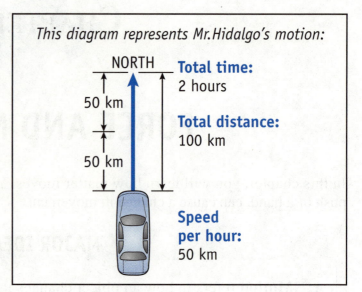

Mr. Hidalgo's motion can also be represented by the following graphs:

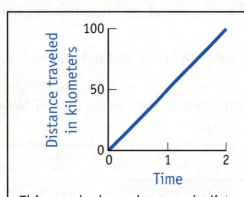

This graph shows how much distance Mr. Hidalgo has traveled. After 1 hour, he has traveled 50 km. After two hours, he has traveled 100 km.

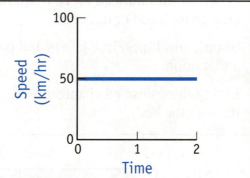

This graph shows Mr. Hidalgo's speed. For the entire trip of two hours, he travels at a constant speed of 50 km per hour.

APPLYING WHAT YOU HAVE LEARNED

♦ Mrs. Jones is traveling on a southbound train to visit her daughter, Mary. Mary lives three hours away by train. The train is traveling 100 km per hour. Demonstrate that you understand how to describe motion by filling in the following blanks and then making a graph representing her trip.

Time: _____ Direction: _____

Distance: _____ Speed: _____

FORCE AND MOTION

If a soccer ball is resting on the ground, you can kick it to make it move. This kick applies force to the ball. A **force** is anything that can put matter in motion. In this case, your kick is the force that causes the soccer ball to move.

The ball will start rolling, but it eventually slows down. Scientists once thought all objects would naturally return to a state of rest. If they were moving, they would eventually come to a stop.

In the 1600s, **Sir Isaac Newton** reached a different conclusion. He believed that objects stop rolling because of friction with the ground. **Friction** is the force caused by the rubbing of two surfaces against each other. It resists movement. Newton said that without friction, the soccer ball you kicked would keep rolling forever until some other force stopped it. In fact, that is what would happen if the ball were moving in outer space. This conclusion helped Newton develop his famous **laws of motion**.

A foot applies force to a soccer ball.

BALANCED FORCES

Sometimes an object is subject to more than one force at the same time. If all the forces acting on the object are perfectly balanced against one another, they will not cause the object to move, or affect its motion if it is already moving.

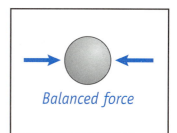

Balanced force

UNBALANCED FORCES

If the forces are not perfectly balanced, however, more force will act on one side of the object than on the other sides. This **unbalanced force** will change the speed or direction of the object. The object will speed up, slow down, change its direction. A change in motion is known as **acceleration**. An unbalanced force will also put into motion an object that is at rest.

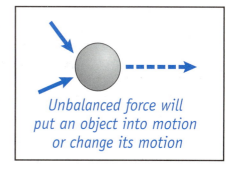

Unbalanced force will put an object into motion or change its motion

APPLYING WHAT YOU HAVE LEARNED

✦ When a car moves ahead, the force from the wheels moving the car forward is greater than the opposing force of friction — the tires rubbing against the roadway.

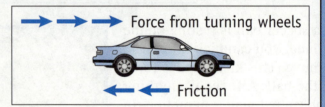

Name two other examples from everyday life in which unbalanced forces lead to changes in an object's speed or direction.

In other words, to increase the speed of an object in motion, you have to give it a "push" in the same direction. To slow it down, you have to give it a "push" in the opposite direction. To get it moving from a state of rest, you have to give it a "push" in the direction you want it to move. How great must this "push" or **force** be to change the speed of the object?

★ If the **mass** of the object is greater, the force must also be *greater*.

★ To make a greater change in **speed**, the force must be *greater*.

Thus, the force required must be ***proportional*** to both the mass of the object and its change in speed. Since **acceleration** means a change in speed, scientists explain this as force equals mass times acceleration, or F = m • a. Force is measured in **newtons** (N).

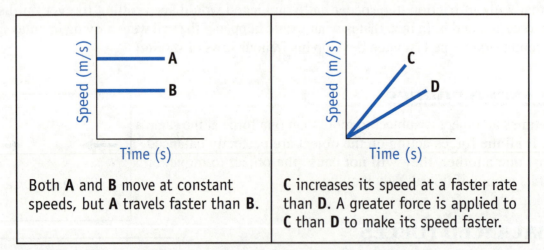

Both **A** and **B** move at constant speeds, but **A** travels faster than **B**.

C increases its speed at a faster rate than **D**. A greater force is applied to **C** than **D** to make its speed faster.

APPLYING WHAT YOU HAVE LEARNED

✦ Dora's parents are driving at 20 miles per hour. Her mother pushes the car's gas pedal to increase its speed. The car takes 2 minutes to increase its speed from 20 to 40 miles per hour. Draw a line graph to show the speed of the car. Label one axis *"Time"* and the other axis *"Speed."*

HOW SIMPLE MACHINES MULTIPLY FORCE

Simple machines change the direction of a force or multiply a force by changing the distance over which it is applied. **Work** is the amount of **force** applied times the **distance** over which it is applied. The amount of work that goes into a machine will equal the amount of work that comes out. The same force can therefore move either a smaller mass over a greater distance or a larger mass over a shorter distance.

Work = force x distance **W = fd**

LEVERS

A **lever** is a simple machine that consists of an **arm** that turns on a **fulcrum**. When you push down on one side of the lever (*input force or effort*), the other side goes up (*output force or load*). A see-saw is a kind of lever.

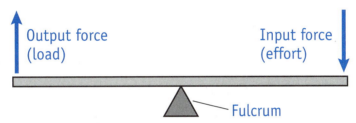

If you move the fulcrum towards one side of the lever, the application of force on the longer side of the lever will create an even greater force on the shorter, opposite side. This is because the force times the distance on each side of the fulcrum must be the same.

In this case, the output force is greater than the input force. Can you explain why?

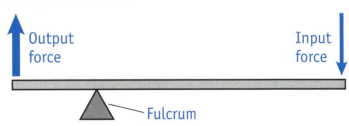

There are different kinds of levers, based on where the fulcrum is placed. For example, the fulcrum could be at one end of the lever, with the output force between the fulcrum and the input force.

APPLYING WHAT YOU HAVE LEARNED

✦ Many of the everyday objects you use are based on the lever principle. For example, a children's seesaw, a hammer, and a hinged door — all have an ***input force*** (*effort*), ***output force*** (*load*), and a ***fulcrum***. Name two other objects that are forms of levers.

1. _____ 2. _____

INCLINED PLANES

An **inclined plane**, or ramp, increases the distance over which force is applied. This allows a smaller amount of force to move an object. It is easier to move an object a little at a time rather than all at once. For example, to push an object up a ramp takes less force than to lift the object straight upwards. The object travels a greater total distance on the ramp, but less force is needed to raise it.

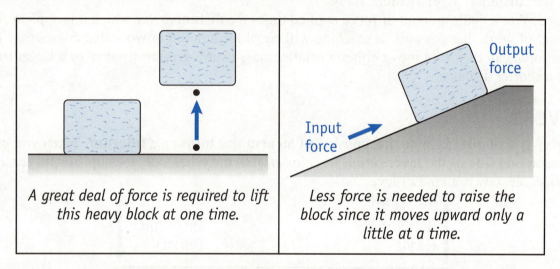

A great deal of force is required to lift this heavy block at one time.

Less force is needed to raise the block since it moves upward only a little at a time.

★ **A screw** is a form of inclined plane that is wrapped around a cylinder. The force applied in turning the screw is multiplied because the screw is an inclined plane. When you turn the screw, it moves forward only a little at a time but with enough force to move through wood.

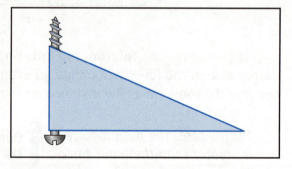

★ **A wedge** is made of two inclined planes attached back-to-back, and can be used to split things. An axe head is an example of a wedge.

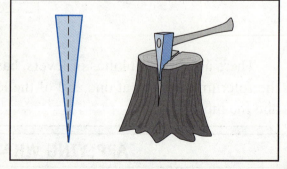

PULLEYS

A **pulley** is another type of simple machine that makes moving heavier objects easier. You use a pulley when you raise a flag up a flagpole or hoist a sail on a sailboat. In both examples, a pulley is made of a rope or string wrapped around the rim of a wheel. You pull the rope to raise something in the opposite direction. The pulley changes the direction of the force applied to it. This allows you to pull down with gravity to lift something upwards.

If you pull the rope at an angle, you have to pull the rope a longer distance to raise the load a lesser distance. In this case, the amount of force is increased.

Two or more pulleys can act almost like levers. The two pulleys can be put together so that as you pull the rope, the load is lifted only half the same distance (see the illustration on the right below). The force lifts the weight a shorter distance than the length that the rope is pulled. In this instance, the amount of force is doubled because it is applied over only half the distance it is pulled. By attaching a series of pulleys together, you can multiply your applied force even more.

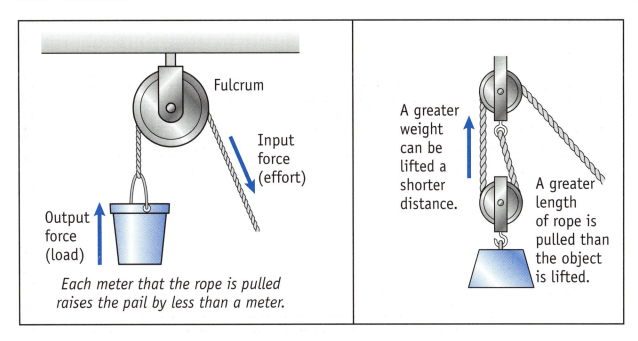

WHEELS AND GEARS

A **wheel** is a circle that is attached to a narrow cylinder or rod, known as the **axle**. When you turn the axle, the wheel turns a greater distance, with less force.

Grooves or teeth on one wheel can be fitted into the groves or teeth of another wheel so that one wheel will turn the other. This is called a **gear** and can be used to transfer motion. This is the guiding principle that allows bicycles, cars, and watches to function.

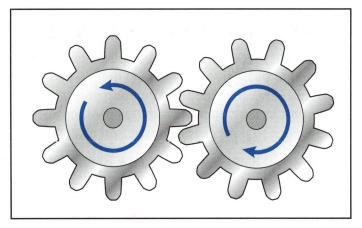

If the size of one of the gears is smaller than the other gear, the force will be increased. If a gear is larger, its teeth will move with less force but at a greater speed than the smaller gear.

APPLYING WHAT YOU HAVE LEARNED

✦ List three different machines you have seen used to pull or lift objects. Indicate whether each machine is a lever, inclined plane or pulley.

Type of Machine	Lever, Inclined Plane, or Pulley?
1. _____	1. _____
2. _____	2. _____
3. _____	3. _____

✦ Force is measured in **newtons** (N). Suppose that a lever has a fulcrum that is 1 meter from side A and 3 meters from side B. How many newtons would be needed on side A to balance a force of 5 newtons on side B?

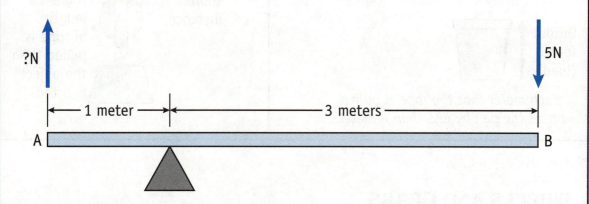

FORCE AND LIVING ORGANISMS

Mechanical force also affects processes in living organisms. For example, mechanical force is needed to push blood around the human body. The heart acts as a pump. Its chambers contract, squeezing the blood and creating pressure that pushes the blood through the body. Force is also required by plants. Seedlings push upwards against gravity. Their root tips are shaped like a wedge to push down into the soil as they grow.

You will learn more about how living organisms use mechanical force in many of their basic processes in Chapter 10 dealing with organisms and organ systems.

CHAPTER 7: FORCE AND MOTION

WHAT YOU SHOULD KNOW

★ You should know that motion can be measured. **Speed** is the average distance traveled by a moving object in a given unit of time, such as m/sec or km/hr. **Direction** describes where the moving object is going.

★ The application of an **unbalanced force** to an object will cause it to change its motion. If an object is at rest, it will start to move in the direction of the force. If it is already moving, the force will cause it to change its direction or speed.

★ You should know that **simple machines**, like levers and pulleys, can change the relationship between force and motion.

★ **Mechanical forces** cause movement in living organisms: for example the movement of blood through the body or the movement of seedlings through the ground.

CHAPTER STUDY CARDS

Motion

★ **Distance.** Total length moved by an object (often in meters).

★ **Speed.** Average distance traveled by a moving object in a unit of time, such as m/sec or miles per hour.

$$\text{Speed} = \frac{\text{distance}}{\text{time}} \quad S = \frac{d}{t}$$

★ **Acceleration.** Change in the motion of an object.

Force

★ **Force.** "Push" or "pull" causes an object to change its motion; measured in newtons.

★ Amount of force needed to change the motion of a body is proportional to the body's mass and speed of change. **F = ma**

★ **Balanced vs. Unbalanced Forces.** Balanced forces have no effect on an object's motion. A force or group of forces that push more on one side than the other is "unbalanced" and causes a change in movement.

Lever

★ A **lever** allows a force applied on the longer side of the lever to create a greater force on the shorter side of the lever.

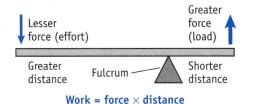

Work = force × distance

Pulleys

★ A **pulley** can change the direction of a force. Multiple pulleys allow a person to pull a rope and raise a weight a shorter distance than the rope is pulled. This increases the amount of force.

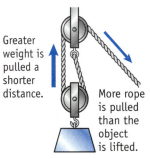

Greater weight is pulled a shorter distance.

More rope is pulled than the object is lifted.

CHECKING YOUR UNDERSTANDING

1. Which of the following graphs best represents a ball moving in outer space at a constant speed of 2 m/sec?

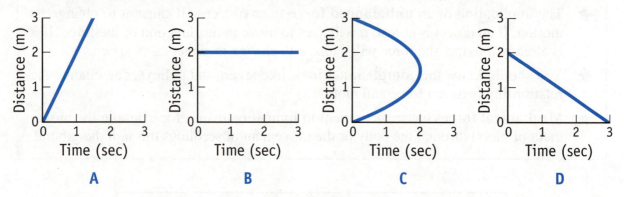

This question asks you to examine four graphs and to choose the one that best represents a ball moving at a constant speed in outer space. You should understand that the **X axis** shows the number of seconds the ball has been moving, while the **Y axis** shows the distance it has traveled in that time. Choice B shows the same distance after 1, 2, and 3 seconds, and is incorrect. Choice D shows no distance traveled after 3 seconds. Choice C makes no sense at all. Only choice A shows the correct distance traveled after 1 second, and $1\frac{1}{2}$ seconds if the ball moves 2 meters every second.

Now try answering some additional questions about force and motion.

2. A caterpillar crawls from point A to point B in 5 seconds. What is the caterpillar's speed? [Use a ruler to measure the distance in centimeters.]

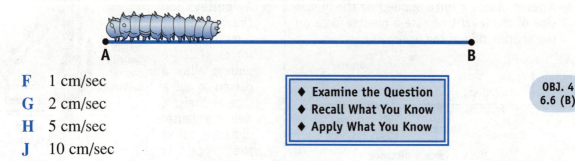

F 1 cm/sec
G 2 cm/sec
H 5 cm/sec
J 10 cm/sec

♦ Examine the Question
♦ Recall What You Know
♦ Apply What You Know

OBJ. 4
6.6 (B)

3 The graph on the right shows the distance traveled by a truck over a period of 80 minutes. During which segment of the graph was the truck going at the fastest speed?

A Line F
B Line G
C Line H
D Line I

OBJ. 4
6.6 (B)

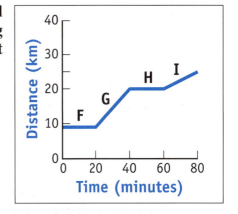

4 The diagram to the right shows a large net force being constantly applied to a basketball. Students are measuring the motion of the ball as part of an experiment in science class. What is one measurement they could take to describe the motion of the basketball?

F The distance it travels every 5 seconds
G The weight of the basketball
H The force applied to the ball
J The height that the ball can bounce

OBJ. 4
6.6 (B)

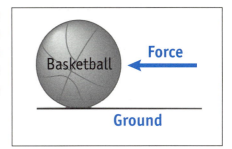

5 A girl leaves her science class and walks 10 meters north to a drinking fountain. Then she turns around and walks 30 meters south to an art classroom. What is the girl's total distance and direction from the science to the art classroom?

A 20 m south
B 40 m south
C 20 m north
D 40 m north

♦ Examine the Question
♦ Recall What You Know
♦ Apply What You Know

OBJ. 4
6.6 (B)

6 Force is measured in newtons (N) — the amount of force needed to speed up the movement of 1 kilogram by 1 m/s^2. Two forces are applied to a ball at rest. One force is twice that of the other. Which diagram best illustrates the motion of the ball?

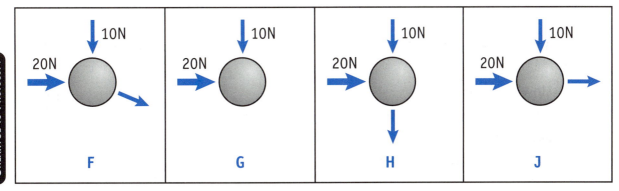

7 Based on the graph, which train traveled the fastest?

A Train A
B Train B
C Train C
D Train D

OBJ. 4
6.6 (B)

♦ Examine the Question
♦ Recall What You Know
♦ Apply What You Know

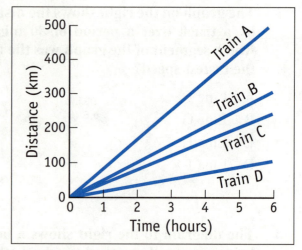

8 Which simple machine could best be used by someone trying to lift a heavy object up to a window on the second story of a house?

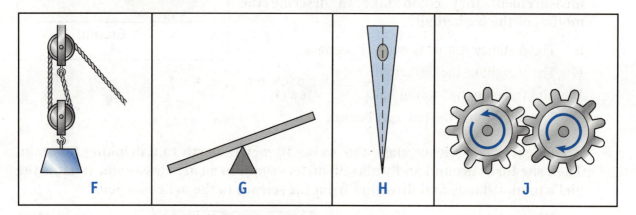

9 Which of the objects shown on the chart would require the most force to achieve an equal increase in speed on a frictionless surface?

Egg	Basketball	Metal Plate	Glass Vase
10 g	100 g	200 g	400g

A egg
B basketball
C metal plate
D glass vase

OBJ. 4
7.6 (A)

10 If all the forces acting on a moving object are perfectly balanced, then the object will —

F slow down and stop
G change the direction of its motion
H accelerate uniformly
J continue moving at a constant speed

OBJ. 4
8.7 (A)

CHAPTER 7: FORCE AND MOTION   89

11 Use the diagram to the right to answer the following question. Which of the following weights could be placed on the left end of the lever to balance the 40 kilogram weight on the right side?
OBJ. 4
8.7 (A)

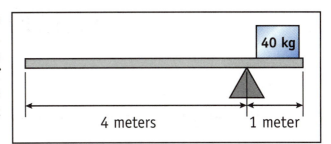

A 5 kilograms
B 10 kilograms
C 40 kilograms
D 160 kilograms

12 A group of bike riders took a 4 hour trip. During the first 3 hours, they traveled at a constant speed of 12 kilometers per hour. During the last hour, they traveled only 10 kilometers. What was the total distance in kilometers traveled by the group during the entire trip? Record and bubble in your answer on the grid below.

OBJ. 4
7.6 (A)

A few questions on the **Middle School TAKS in Science** may ask you to mark your answer on a grid. The griddable format is used to give you the opportunity to provide a numerical answer to a question. Some answers that require a griddable response will ask for a percentage or ask you to measure something with precision. Notice that the grid has a column for decimals. You must record your answer in the columns for the correct place values. When recording answers with whole numbers, you need to add zeros after the decimal. If the answer is a fraction, add a zero in front of the decimal. Answer the question above on the grid form to the right. Write the number of kilometers the bikers rode in the boxes. Then fill in the corresponding circle below each number.

13 A woman bends her arm at the elbow in order to lift a 25 pound weight in her hand. Her arm muscles pull the bones of her forearm upwards.
In this example, her arm acts as a —
A gear
B inclined plane
C lever
D pulley

OBJ. 4
7.6 (A)

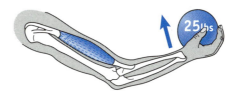

CHAPTER 8

THE NATURE OF ENERGY

In this chapter, you will learn about the nature of energy.

MAJOR IDEAS

A. **Energy** is the ability to do work.

B. Energy can take different forms. **Kinetic energy** is the energy of movement. An object's kinetic energy depends on its mass and speed. Heat energy is a form of kinetic energy, based on the motion of atoms and molecules.

C. **Potential energy** is the energy stored in an object, based on its position or condition.

D. Energy can change from one form to another. Even when energy changes from one form to another, it is always conserved; no energy is lost.

E. **Waves** have energy and can transfer energy through different materials.

TYPES OF ENERGY

As you know, **force** must be applied to an object to change its motion. **Work** is the application of force over distance. **Energy** is the ability to do work. There are many types of energy, including:

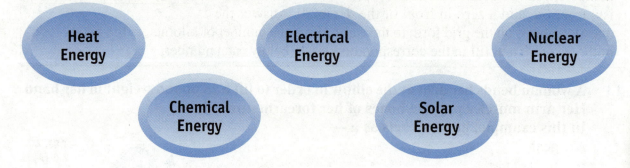

Most forms of energy can be classified as either kinetic energy or potential energy. Let's examine each to see how they differ.

KINETIC ENERGY

Any moving object is able to do work: therefore it has energy. This energy of motion is known as **kinetic energy**. When you walk, run or jump, your body exhibits kinetic energy. When water falls, it has kinetic energy.

The amount of kinetic energy a moving object has depends on two factors: mass and speed. The greater an object's speed, the greater its kinetic energy; the greater its mass, the greater its kinetic energy.

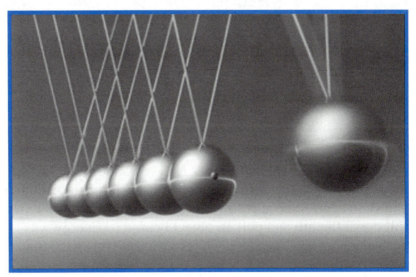

Because this ball is moving, it has kinetic energy and can use this energy to move other objects.

HEAT ENERGY

Heat energy is also a form of kinetic energy. The **kinetic theory** explains that what we feel as heat is actually caused by the random motion and vibration of atoms and molecules in substances. The heat created by this movement is a form of kinetic energy. **Temperature** is a measure of the speed of movement of these atoms and molecules: it is the average kinetic energy of the particles in an object. An increase in temperature represents an increase in molecular motion. The higher the temperature of an object, the greater the kinetic energy of its atoms and molecules.

APPLYING WHAT YOU HAVE LEARNED

◆ Use the kinetic theory to explain what happens to ice as it melts.

POTENTIAL ENERGY

Potential energy is the energy that an object is able to *store* because of its position or condition. Think of a metal spring. As the spring is pushed down, potential energy is stored in its coils. Once you let go of the spring, it bounces back up, converting its potential energy into kinetic energy.

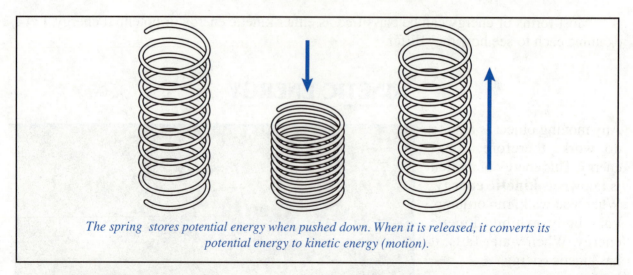

The spring stores potential energy when pushed down. When it is released, it converts its potential energy to kinetic energy (motion).

One type of potential energy is based on gravity. An object that works against gravity to move into its position actually stores energy. Its position on Earth's surface gives the object the ability to exert force and do work. For example, if a truck is pushed to the top of a hill, it can roll down the hill and pull an object behind it. Energy was "stored" in the truck when it was pushed up the hill. The amount of potential energy in the truck depends on its weight and height above Earth's surface.

Chemical energy is another form of potential energy. Some types of molecules store energy in the bonds formed by their shared electrons. When these bonds are broken, their energy is released.

OTHER FORMS OF ENERGY

ELECTRICITY

Another form of energy is electricity. Electricity is created by the movement of electrons. These electrons carry negative electrical charges, which can travel through some substances and can even move from one substance to another. Electricity can flow in a circuit and can produce heat, light, sound and magnetism.

NUCLEAR ENERGY

Nuclear energy is another important form of energy. When large nuclei split apart, they release free neutrons and energy. Small amounts of matter are converted into immense quantities of energy.

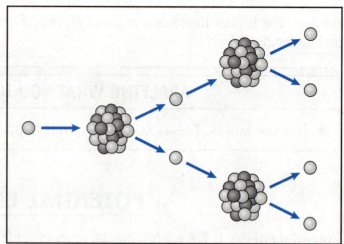

A nuclear chain reaction releases the energy stored in the nuclei of atoms.

Joining the nuclei of smaller atoms together can also release stored nuclear energy. The sun, for example, fuses together the nuclei of hydrogen atoms into helium to produce its energy.

APPLYING WHAT YOU HAVE LEARNED

+ How does a bucket resting on the side of a table have potential energy? Explain your answer.
+ What is nuclear energy? Explain where its power comes from.

THE TRANSFORMATION OF ENERGY

One important quality of energy is its ability to change its form. For example, potential energy can change into kinetic energy, and kinetic energy can change into potential energy. This ability to change form allows energy to move through a substance or even to transfer from one substance to another.

As a roller coaster car loses height, it gains speed, Potential energy is converted to kinetic energy. As it gains height, the car loses speed: the kinetic energy is transformed into potential energy.

Think about a roller coaster ride. The roller coaster car is brought to the top of a hill. It now has stored potential energy. As it starts to move downward, the roller coaster car's potential energy decreases. At the same time, its kinetic energy increases. This is because the car's kinetic energy equals its mass times its speed. As it rolls downward, its speed increases. This kinetic energy then pushes the roller coaster car up the next incline. As it moves upward, its potential energy is again increased, while its kinetic energy decreases as the car slows down.

CHEMICAL REACTIONS

Other kinds of energy, besides mechanical energy, can also change their form. For example, chemical energy often converts into heat energy when a chemical reaction occurs; other chemical reactions may absorb energy.

In photosynthesis, plants take light from the sun and convert this into chemical energy. Other chemical reactions that require energy take place when substances are heated, or absorb heat energy from the surrounding environment.

Chemical reactions that release energy, like the burning of wood, often transform chemical energy into heat energy that is then released into the environment.

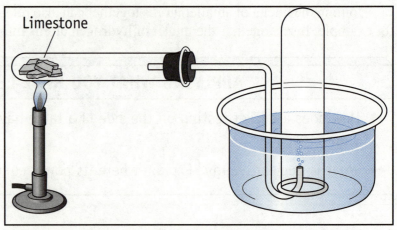

Some chemical reactions require heat. Limestone can break down into lime and carbon dioxide gas, but only if the limestone is heated.

POWER-DRIVEN MACHINES

To meet their needs, humans often make use of the ability of energy to change its form. For example, a hydroelectric plant uses power from falling water to turn machines that make electricity. This electricity is carried by wires into homes and businesses, where it is converted into heat and light.

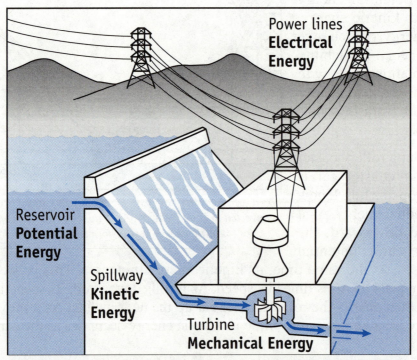

A hydroelectric power plant changes energy from one form to another. This illustration shows a hydroelectric plant using the kinetic energy of falling water to turn a turbine, converting some of this kinetic energy into electrical energy.

The Internal Combustion Engine. The internal combustion engine found in automobiles uses the chemical energy stored in fossil fuel (*gasoline*). It releases this energy by burning the fuel, causing an explosion of gas and heat. The expansion caused by this heat energy forces a piston in the engine to go up and down. In this example, the engine has turned chemical energy into heat energy and then into mechanical energy.

The Conservation of Energy. Although energy can change from one form to another, it cannot be created or destroyed. This principle is known as the **conservation of energy**.

APPLYING WHAT YOU HAVE LEARNED

✦ List two examples from everyday life in which energy changes its form.

✦ Explain what is meant by the "conservation of energy."

WAVES OF ENERGY

Some forms of energy spread in special patterns known as waves. A **wave** is a vibration or disturbance that carries energy through matter or space. The waves move away from the source as surface ripples, circles, or spheres, like the ripples from a stone thrown into a pond. Seismic waves, water waves, sound waves and light waves all transfer energy in this form.

There are two main types of waves: **mechanical** and **electromagnetic**.

Mechanical   **Electromagnetic**

MECHANICAL WAVES

Mechanical waves — such as water waves or sound waves — always move through some form of matter. Some force usually sets the particles of matter in motion. The particles in the matter move or vibrate and then pass this energy on to neighboring particles. Sound, for example, is made by a vibrating object. The object starts vibrations in the air which reach our ears. The type of matter that the waves travel through, known as the **medium**, often affects the properties of the wave. For example, **sound waves** generally travel faster in solids or liquids than in the air. This occurs because sound waves are caused by the vibrations of particles. Since particles are closer together in solids and liquids than in air, sound vibrations are able to travel faster through them. Ocean waves are another example of mechanical waves. The water in a wave rises and falls. It then passes its energy to neighboring water, which also rises and falls.

ELECTROMAGNETIC WAVES

Not all waves travel through matter. **Electromagnetic waves**, such as radio waves or light, do not require a medium. They can travel through some forms of matter, but they can also travel through a **vacuum** (*empty space*). They do not require particles of matter to carry their energy. You can think of these waves as bursts of pure energy. Solar energy from the Sun radiates to Earth in this way.

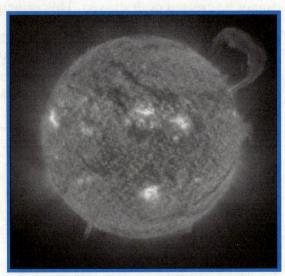

The Sun radiates energy to Earth.

Scientists often use special terms to describe the characteristics of waves:

DESCRIBING WAVES

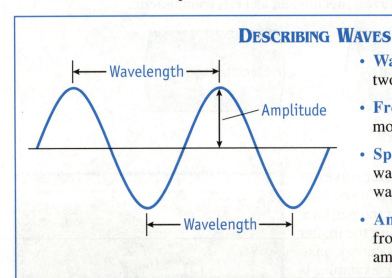

- **Wave length.** The distance between two similar parts of a wave.
- **Frequency.** How many waves move in a given period of time.
- **Speed of Wave.** The speed of a wave equals its frequency times its wavelength.
- **Amplitude.** The height of a wave from its midpoint, showing its amount of energy.

Because sound and light are made up of waves, their speed can be measured. Light travels very fast — 300,000,000 meters in one second! Sound travels much more slowly. Its speed is about 343 meters per second (at sea level at 20° C). That is why you can see a lightning flash before you hear its thunder.

THE ELECTROMAGNETIC SPECTRUM

Light waves, radio waves and all other types of electromagnetic radiation are identified by the **electromagnetic spectrum**, which is made up of visible light waves and other forms of electromagnetic radiation we cannot see. All forms of electromagnetic radiation travel at the same speed. However, their frequencies and wavelengths differ.

CHAPTER 8: THE NATURE OF ENERGY   97

The electromagnetic spectrum classifies these different types of waves by **frequency** — the number of waves that occur in a given period of time. The lowest frequencies on the electromagnetic spectrum have the longest wavelengths. As electromagnetic waves increase in frequency, their wavelengths become shorter.

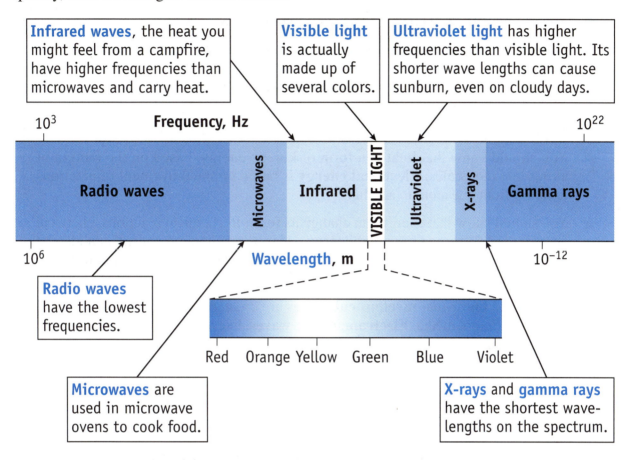

APPLYING WHAT YOU HAVE LEARNED

✦ List four ways in which electromagnetic radiation affects your life. The first example has been done for you:

Type of Electromagnetic Radiation	Impact on Your Life
1. Ultraviolet light	1. Leads to sunburn
2.	2.
3.	3.
4.	4.

WHAT YOU SHOULD KNOW

★ You should know that **energy** is the ability to do work.

★ You should know that energy comes in different forms.
  - **Kinetic energy** is the energy of motion. It is based on a moving object's mass and speed.
  - Energy can also be stored as **potential energy**. One form of potential energy is the energy stored in an object because of its position on Earth.
  - Energy can also be stored in chemical bonds.

★ You should know that **heat** is a form of kinetic energy, based on the motion of atoms and molecules. **Electrical energy** is based on the movement of electrons jumping from one atom to another.

★ You should know that energy can change its form. In chemical reactions, chemical energy is often turned into heat energy. Heat or mechanical energy can also be turned into electrical energy. The total amount of energy is always conserved, even when the form of energy is changed.

★ You should know that **mechanical waves** are created by a force and can travel through different media. **Electromagnetic waves**, like sunlight, can travel through either matter or across a vacuum.

CHAPTER STUDY CARDS

Forms of Energy

★ **Energy.** The ability to do work.
★ **Kinetic Energy.** Energy of motion based on the mass and speed of the moving object.
★ **Heat Energy.** A form of kinetic energy based on the vibrations and movements of atoms and molecules.
★ **Potential Energy.** Stored energy.
  - One form of potential energy is based on an object's position on Earth and the force exerted by gravity.
  - **Chemical Bonds** are another form of potential energy.

Transformation of Energy

Energy can change from one form to another. For example:

★ Kinetic energy can turn into potential energy and back again.
★ Heat energy can be used to create electrical energy. Electrical energy can be changed into heat energy.
★ **Law of Conservation.** Energy can change its form, but its total quantity is always conserved.

Other Forms of Energy	Waves
★ **Electricity.** Electricity is created by the movement of electrons. ★ **Nuclear Energy.** When large nuclei split apart, they release energy. Small amounts of matter can be converted to huge quantities of energy. When joined together, the nuclei of smaller atoms can also release nuclear energy.	★ **Mechanical Waves.** Seismic, water, or sound waves — pass through a medium; particles of the medium pass along the energy of the wave. ★ **Electromagnetic Radiation.** These types of wave can pass through some forms of matter but do not require it; they can also pass through a vacuum or outer space. Electromagnetic radiation includes invisible waves as well as visible light.

CHECKING YOUR UNDERSTANDING

1 A cement block is dropped from a helicopter. The block falls 100 feet. Which statement accurately describes the falling cement block?

   A   Its potential energy decreases as its kinetic energy increases.
   B   Its potential energy increases as its kinetic energy decreases.
   C   Its potential energy is unchanged as its kinetic energy decreases.
   D   Its potential energy is unchanged as its kinetic energy increases.

OBJ. 4
7.8 (A)

 HINT This question tests your understanding of potential and kinetic energy. Recall what you know about both forms of energy: then apply this knowledge to the question. As the block falls, its speed increases but its height decreases. Its kinetic energy therefore increases while its potential energy decreases. Thus, the correct answer can only be choice A.

Now try answering some questions on your own about the nature of energy.

2 What happens when a substance is heated?

   F   Its lighter particles clump together.
   G   The bonds within its atoms break down.
   H   The molecules in the substance move faster.
   J   The molecules of the substance slow down.

OBJ. 4
7.8 (A)

♦ Examine the Question
♦ Recall What You Know
♦ Apply What You Know

3   The diagram to the right shows a spring between two carts at rest on a frictionless surface. When the spring is released, the two carts move in opposite directions. What will occur when the spring is released?

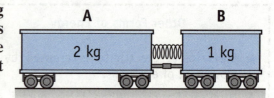

OBJ. 4
7.8 (A)

A   The kinetic energy of cart A will equal that of cart B.
B   The kinetic energy of cart A will equal the potential energy of cart B.
C   The kinetic energy of cart A will increase the potential energy as the spiral decreases.
D   The kinetic energy of cart A will be less than the potential energy of cart B.

4   Which graph represents the relationship between the potential energy of an object based on gravity and its height above the surface of Earth?

OBJ. 4
7.8 (A)

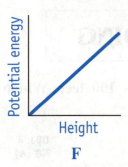
F

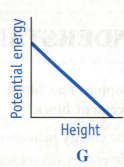
G

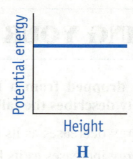
H

J

5   Scientists are preparing to launch a satellite into outer space. Which form of energy will the scientists be able to use to communicate with that satellite?

A   sound waves
B   radio waves
C   sonar
D   electricity

♦ Examine the Question
♦ Recall What You Know
♦ Apply What You Know

OBJ. 4
8.7 (B)

6   Which of the following is designed to change electrical energy into light energy?

OBJ. 4
6.9 (A)

F

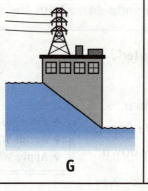
G

H

J

7. Sound travels at the speed of 343 meters per second. A tree crashes in the forest. How many seconds will it take for an observer 1,372 meters away to hear the crash of the tree? Record and bubble in your answer on the grid below.

♦ Examine the Question
♦ Recall What You Know
♦ Apply What You Know

OBJ. 4
8.7 (B)

8. Falling water runs through a hydroelectric power plant. The falling water turns giant turbines in the plant, which produce electricity. The electricity is carried by wires into the homes of people living in Dallas, Texas. There, it powers electric heaters. Which of the following best identifies the energy transformations that have taken place?

F   kinetic energy ⟶ electrical energy ⟶ heat energy
G   electrical energy ⟶ kinetic energy ⟶ potential energy
H   potential energy ⟶ heat energy ⟶ electrical energy
J   heat energy ⟶ electrical energy ⟶ chemical energy

OBJ. 4
6.9 (A)

9. Keesha mows her neighbor's lawn with a gasoline-powered lawn mower. Which best describes the transformation of energy that occurs?

A   chemical energy ⟶ kinetic energy ⟶ thermal energy
B   potential energy ⟶ kinetic energy ⟶ chemical energy
C   electrical energy ⟶ chemical energy ⟶ kinetic energy
D   thermal energy ⟶ kinetic energy ⟶ nuclear energy

OBJ. 4
6.9 (A)

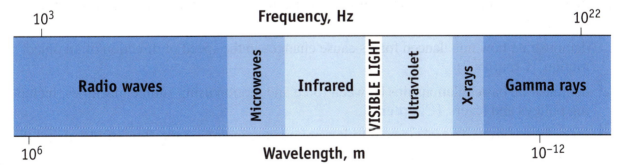

10. Based on the chart, which type of electromagnetic radiation has a longer wave length than microwaves?

F   Gamma rays
G   Radio waves
H   Ultraviolet light
J   X-rays

OBJ. 4
8.7 (B)

11 Wavelengths are measured in meters and frequency (how many waves per second) is measured in hertz. The diagram below shows the frequency and wavelength of various types of electromagnetic radiation.

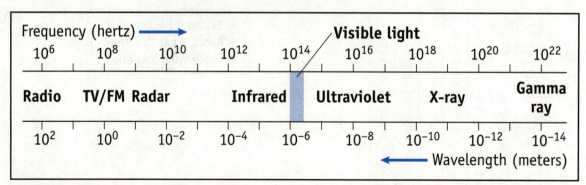

Which type of electromagnetic radiation has a wavelength of approximately 10^4 meter and a frequency of 10^{12} hertz?

A radio
B infrared
C X-ray
D gamma rays

OBJ. 4
8.7 (B)

CHECKLIST OF OBJECTIVES IN THIS UNIT

Directions. Place a check mark (✔) next to those benchmarks you understand. If you have trouble remembering information connected with one of the benchmarks, review the chapter indicated in the brackets for the items you do not recall. **You should be able to:**

❑ demonstrate that changes in motion can be measured and represented by graphs. **[Chapter 7]**

❑ demonstrate how unbalanced forces cause changes in the speed or direction of an object's motion. **[Chapter 7]**

❑ demonstrate basic relationships between force and motion using simple machines, including pulleys and levers. **[Chapter 7]**

❑ relate forces to basic processes in living organisms, including the flow of blood and the emergence of seedlings **[Chapter 7]**

❑ illustrate examples of potential and kinetic energy in everyday life, such as objects at rest, and falling water **[Chapter 8]**

❑ identify energy transformations occurring during the production of energy for human use such as electrical energy into heat energy or heat energy into electrical energy. **[Chapter 8]**

❑ recognize that waves are generated and can travel through different materials *(media)*. **[Chapter 8]**

UNIT 5

LIFE SCIENCES

In this unit, you will review what you need to know for the **Middle School TAKS in Science** about the life sciences — the study of all living things.

You will learn how the cell is the basic unit of life. You will also explore human organ systems, heredity, and evolution and study how living things interact with their environment.

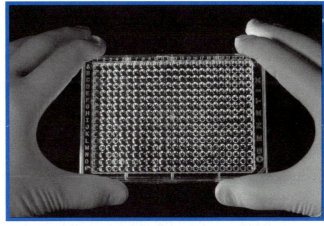

A tray containing human genome DNA

★ **Chapter 9: The Cell**
In this chapter, you will learn about cell theory and how to identify different types of cells and their parts. In addition, you will study cellular processes, including how cells use energy and matter, and how cells grow and divide.

★ **Chapter 10: Organisms and Organ Systems**
In this chapter, you will learn how cells form organs and how groups of organs form organisms. You will also learn about the organs of the human body.

★ **Chapter 11: Heredity and Evolution**
In this chapter, you will learn how some characteristics are inherited. You will also learn how environmental changes affect the survival of organisms.

★ **Chapter 12: Ecosystems**
In this chapter, you will learn about various types of ecosystems, how ecosystems change over time, how energy and matter flow through an ecosystem, and how human activities affect ecosystems.

CHAPTER 9

THE CELL

This chapter examines cells and cellular processes.

MAJOR IDEAS

A. The **cell** is the basic unit of all living things. All existing cells come from pre-existing cells.

B. All cells carry on functions to live, grow, and reproduce. Cells have specialized structures to carry on these processes.

C. Plant cells convert the energy of sunlight into chemical energy through **photosynthesis**.

WHAT IS A CELL?

The **cell** is the smallest unit of matter that is able to carry on all the processes of life. The discovery of **cells** came about after the invention of the microscope. In 1665, the English scientist Robert Hooke (1635-1703) identified the first cells when he looked at a piece of cork through a microscope.

We now know that all living cells share certain common characteristics:

★ They are surrounded by a **cell membrane**.

★ They contain hereditary material in the form of **DNA** (*a large organic molecule*), which they receive from a pre-existing cell or cells. DNA molecules provide the blueprint for how each cell operates.

★ They are mainly composed of fluid, which contains chemicals and structures that allow them to live, grow and reproduce. This fluid and its structures are known as **cytoplasm**.

★ They are composed of a small number of key chemical elements — *carbon, hydrogen, oxygen, nitrogen, phosphorus,* and *sulfur*.

Modern **cell theory** consists of three main parts:

> ★ The cell is the basic unit of structure and function of all living organisms.
>
> ★ All living things are made up of one or more cells, which can be of various types.
>
> ★ All living cells come from the reproduction of pre-existing cells.

CELL STRUCTURES

There are many different types of cells. All cells have specialized structures that help them to live, grow and reproduce.

Bacteria. Bacteria are the simplest types of cells. They have a cell wall and cell membrane covering the cell. Their DNA is just bunched up like a curled rubber band inside the cell.

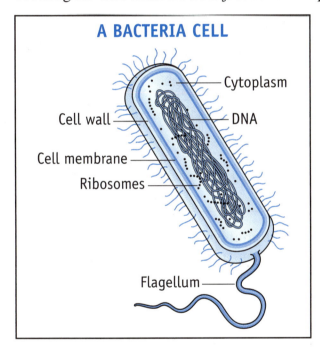

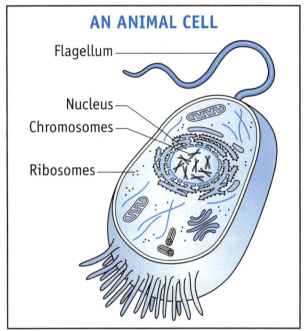

Other Cells. Other cells are more complex. They have the following features:

★ **Cell Membrane.** Each cell is covered by a thin membrane of specialized molecules. The cell membrane holds the cell together. It also controls what enters and leaves the cell.

★ **Nucleus.** A large nucleus near the center of the cell holds the cell's DNA, its hereditary information. The nucleus is covered by its own double membrane.

- **Other Structures**. Other special structures exist in the cell's cytoplasm. These structures are uniquely designed to help the cell perform its essential functions. Some of these structures help move nutrients, gases, water and other particles around the cell. Other structures help the cell break down sugars in the process known as **cellular respiration**. Still others, like the **ribosomes**, help the cell make proteins. **Lysosomes** dissolve wastes.

PLANT CELLS

Plant cells have several additional structures that make them different from animal cells. These features are important in helping the plant to function.

- **Cell Walls**. The cell membranes of plants are covered by a rigid, outer cell wall that supports and protects the plant. A plant's cell walls have pores that allow it to interact with the environment.

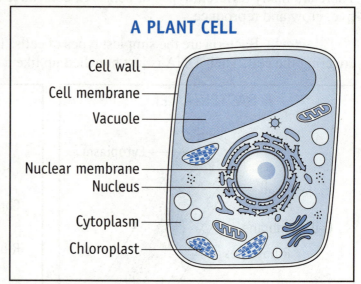

- **Vacuoles**. These are large fluid-filled "bubbles" within plant cells that store and hold water, enzymes and waste. These vacuoles can be quite large, taking up to 90 percent of a plant cell's volume.

- **Chloroplasts**. Plant cells have special structures that allow plants to conduct **photosynthesis** — the conversion of sunlight into chemical energy. Chloroplasts contain a large amount of pigment, often green, that is responsible for giving plants their distinctive color.

HOW ANIMAL CELLS DIFFER FROM PLANT CELLS

Animal Cells	Plant Cells
• Can move about	• Are confined by cell walls
• Must consume other organisms for food	• Conduct photosynthesis (in their chloroplasts)
• Have lysosomes—special structures to dissolve wastes and other products	• Have large vacuoles for enzymes and waste

---

APPLYING WHAT YOU HAVE LEARNED

✦ Make your own drawing of an animal cell and a plant cell. Label their parts.

CELLULAR FUNCTIONS

In order to live, grow and reproduce, all cells must carry on certain functions.

MAINTAIN EQUILIBRIUM

Cells and all living things must maintain stable internal conditions in order to survive. This stability is known as **equilibrium**. For example, a cell must control how much water it has, its temperature, and the speed of its chemical reactions.

THE MOVEMENT OF PARTICLES IN AND OUT OF CELLS

One way a cell maintains its equilibrium is by controlling what enters and leaves it. In order to maintain oxygen, food and water, cells must be able to admit molecules and other particles. They must also be able to reject wastes, excess water, and foreign organisms.

The cell membrane is made up of special molecules that can allow molecules to enter. Some molecules enter through random movement. For example, water molecules will enter the cell randomly if there is more water outside the cell than inside it.

A cell can also use its own energy to surround a particle and ingest it. The particle is surrounded by the cell membrane, pinched off in a membrane-covered sac and then brought into the cell.

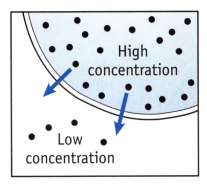

Some molecules enter or leave cells through random motion, going from areas of high concentration to areas of low concentration.

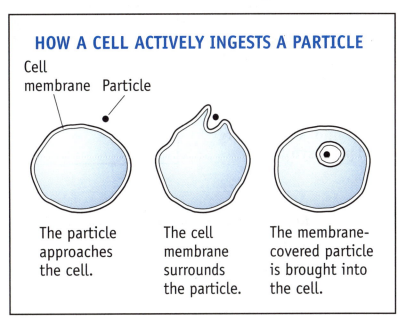

A cell can also expel particles. Structures in the cell wrap the particle in a membrane and then move the particle outside the cell.

TURGOR PRESSURE

The movement of water into a plant cell causes it to swell. The cell membrane presses against the plant's cell wall. **Turgor pressure** is the pressure of water molecules against the cell wall. As turgor pressure increases, more water cells leave the cell than enter it. As turgor pressure decreases, more water cells enter the cell than leave. This **feedback mechanism** helps the plant cell maintain its equilibrium. You will learn more about feedback mechanisms in the next chapter.

HOW CELLS GAIN AND RELEASE ENERGY

All forms of life require energy to carry out the functions of life. Plants obtain their energy from the Sun through photosynthesis. Other living things acquire energy by eating plants or by eating other living things that eat plants. This gives them organic compounds with stored energy.

PHOTOSYNTHESIS

Plants obtain energy from radiant sunlight and store it for later use. Energy from the Sun travels through space to Earth's surface. A green plant needs only a few seconds to capture the energy in sunlight, process it, and store it in the form of a chemical bond. Plants use the energy from sunlight to create glucose, a form of sugar, out of carbon dioxide (CO_2) and water (H_2O). Plants also release oxygen as a by-product of photosynthesis. The process of converting light energy into stored energy is called **photosynthesis**.

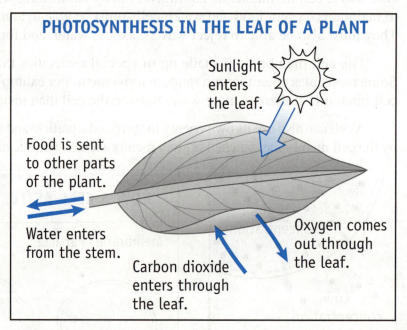

PHOTOSYNTHESIS IN THE LEAF OF A PLANT

Sunlight enters the leaf.

Food is sent to other parts of the plant.

Water enters from the stem.

Carbon dioxide enters through the leaf.

Oxygen comes out through the leaf.

$$6CO_2 + 6H_2O + \text{sunlight} \xrightarrow{\text{chlorophyll}} C_6H_{12}O_6 + 6O_2$$

carbon dioxide   water   glucose   oxygen

CELLULAR RESPIRATION

During **cellular respiration**, cells break down organic compounds to release energy. Cellular respiration is not the same as breathing, but there are similarities: both require oxygen and produce carbon dioxide. Cells use oxygen in a series of reactions to change glucose ($C_6H_{12}O_6$) back into carbon dioxide (CO_2) and water (H_2O). These reactions release the energy stored in the glucose by photosynthesis. In fact, cellular respiration is almost the reverse of photosynthesis.

APPLYING WHAT YOU HAVE LEARNED

◆ Make a chart comparing *photosynthesis* and *cellular respiration*.

HOW CELLS DIVIDE

You know that all living cells come from pre-existing cells. Cells reproduce through the process of cell division. Cell division requires copying of the cell's **DNA** — the hereditary instructions that tell the cell how to function.

CELL DIVISION WITHOUT A NUCLEUS

Bacteria have no distinct nucleus. To divide, the DNA in the cell simply copies itself. Then a new cell wall develops, separating the two new cells. Each new cell is identical to the parent cell.

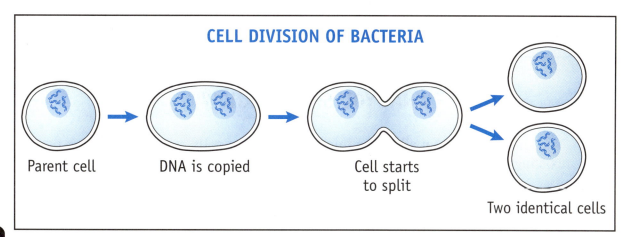

CELL DIVISION WITH A NUCLEUS

In cells with a nucleus, strands of DNA form coils known as **chromosomes** before the cell is going to divide. These chromosomes — containing the cell's DNA — copy themselves first. As the illustration on the next page shows, the membrane separating the nucleus from the rest of the cell then disappears. The chromosomes line up across the middle of the cell and then divide in half.

A new nuclear membrane forms around the chromosomes on each side. The cell membrane then pinches the cell's cytoplasm in two. Two new cells, each identical to the original cell, have formed.

Both cell division without a nucleus and cell division in which the nucleus copies itself are forms of *asexual reproduction*. The offspring are identical to the parent cell.

Sex cells are special cells that are used by plants and animals for *sexual reproduction*. In these forms of life, chromosomes come in pairs. Each pair of chromosomes determines one characteristic.

Sex cells are used to pass these characteristics to offspring. When sex cells divide, each new cell receives only half the chromosomes of the parent. Later, the sex cell will join with a sex cell from another plant or animal to form an offspring with some hereditary characteristics from both parents. You will learn more about heredity and genetics in Chapter 11.

HOW A CELL WITH A NUCLEUS DIVIDES

- Parent cell
- Chromosomes are copied into pairs
- The pairs line up in the middle
- The pairs separate
- Two daughter cells — The cell membrane pinches the cell in two

APPLYING WHAT YOU HAVE LEARNED

♦ Identify three functions that cells carry on to sustain life. Then explain how cells accomplish each function.

WHAT YOU SHOULD KNOW

★ You should know that all living things are made of cells, which come from pre-existing cells.

★ You should know that all cells have cell membranes, cytoplasm, and DNA.

★ You should know that plants convert radiant sunlight into stored chemical energy in a process known as **photosynthesis**.

★ You should know that cells release energy through cellular respiration.

★ You should know that cells divide and multiply by copying their DNA, which contains each cell's hereditary information.

CHAPTER 9: THE CELL   111

CHAPTER STUDY CARDS

Cell Theory

Modern cell theory consists of three parts:

★ **Basic Unit of Living Things.** The cell is the basic unit of all living things.

★ **Living Things Are Made of Cells.** All living things are made up of one or more cells.

★ **Come from Pre-existing Cells.** All living cells come from the reproduction of pre-existing cells.

Cell Structures

Specialized cell structures help cells meet their basic needs.

★ **Cell Membrane** holds cell together and controls what enters and leaves the cell.

★ **Cytoplasm** is the fluid and structures inside the cell membrane.

★ **Nucleus:** All cells more complex than bacteria have a membrane-covered nucleus that holds their DNA.

Cell Functions

★ **Equilibrium.** A cell must maintain stable internal conditions.

★ **Movement of Molecules / Waste.** A cell controls what enters and leaves it.
 • Some molecules enter or leave the cell through random motion.
 • Other particles are ingested or expelled by the cell.

★ **Turgor Pressure.** Pressure of water against a plant cell wall.

★ **Cell Division.** Cells divide by copying their own DNA.

How Cells Gain and Release Energy

All organisms use energy to carry out the functions of life:

★ **Photosynthesis.** Plants use light to convert carbon dioxide and water into glucose and oxygen. Energy is stored in the chemical bonds of glucose molecules.

★ **Cellular Respiration.** Cells break down glucose into usable energy; the reverse of photosynthesis.

CHECKING YOUR UNDERSTANDING

1. A defective cell loses its ability to control the passage of water, food, and waste into and out of the cell. In which structure of the cell is this defect located?

   A The chloroplasts
   B The cell membrane
   C The vacuoles
   D The nucleus

   ♦ Examine the Question
   ♦ Recall What You Know
   ♦ Apply What You Know

   OBJ. 2
   6.10 (B)

HINT To answer this question correctly, you must identify the structure of a cell that controls the movement of water, food and wastes into and out of it. Choice A is incorrect since chloroplasts allow a plant to convert sunlight into energy. From the structures remaining, which one determines what enters and leaves the cell?

Now try answering some additional questions dealing with the topics covered in this chapter — cells, cellular functions, and cell division.

2 One difference between plant and animal cells is that animal cells lack —

   F a nucleus
   G chloroplasts
   H a cell membrane
   J ribosomes

♦ Examine the Question
♦ Recall What You Know
♦ Apply What You Know

OBJ. 2
6.10 (B)

3 The ability of cells to pass on their characteristics to new cells is directly related to the —

   A size of their nucleus
   B thickness of the cell membrane
   C ability of plants to convert solar energy
   D ability of chromosomes to copy themselves

OBJ. 2
6.10 (B)

4 After the division of a sex cell, each new sex cell has —

   F all the chromosomes of the parent
   G none of the chromosomes of the parent
   H half the chromosomes of the parent
   J double the chromosomes of the parent

OBJ. 2
6.10 (B)

5 The diagram to the right represents a process that often occurs in the cells of a plant. What process is illustrated?

   A cellular respiration
   B photosynthesis
   C cell division
   D turgor pressure

OBJ. 2
7.8 (B)

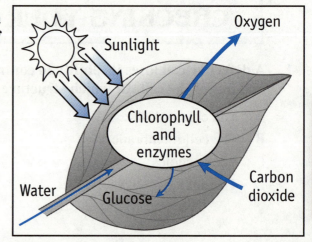

CHAPTER 9: THE CELL   113

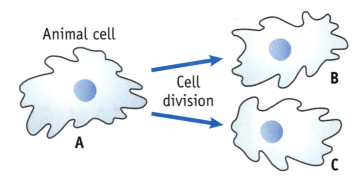

6 Which statement about the cells in the diagram above is correct?

    F   Cell A contains the same DNA as cells B and C.
    G   Cell C has DNA that is only half identical to cell B.
    H   Cells B and C have identical DNA but not the same as Cell A.
    J   Cells A, B, and C contain completely different DNA.

OBJ. 2
6.10 (B)

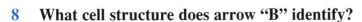

Use the diagram to right to answer questions 7 and 8.

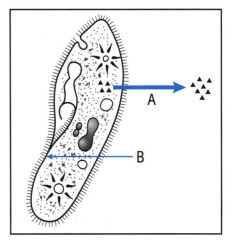

7 This diagram represents a single cell living in a watery environment. If the triangle symbols (▲▲) represent molecules of a specific substance, what process does arrow "A" represent?

    A   Cell division
    B   Cellular respiration
    C   Photosynthesis
    D   Movement of particles

OBJ. 2
8.6 (B)

8 What cell structure does arrow "B" identify?

    F   The nucleus
    G   The cell membrane
    H   The chloroplasts
    J   The vacuoles

OBJ. 2
6.10 (B)

9 "Living things contain units of structure and function that arise from pre-existing units." This statement best summarizes —

    A   cell theory
    B   photosynthesis
    C   equilibrium
    D   cell respiration

♦ Examine the Question
♦ Recall What You Know
♦ Apply What You Know

OBJ. 2
6.10 (B)

CHAPTER 10

ORGANISMS AND ORGAN SYSTEMS

An **organism** is any living thing that can live on its own. In this chapter, you will learn how groups of cells make up some organisms, and how various organ systems work together in the human body.

MAJOR IDEAS

A. A **tissue** is a group of similar cells that work together.

B. An **organ** is a group of two or more tissues that work together to perform a specific function.

C. A **system** has its own properties, which are different from those of its parts.

D. An **organ system** is a group of organs that work together to perform a function in an organism.

E. The human body has several important organ systems that enable it to survive. These include the digestive, respiratory, circulatory, skeletal, muscular, endocrine, and reproductive systems. These organ systems often **interact**.

WHAT IS AN ORGAN SYSTEM?

Many cells are **unicellular organisms**. These cells can live on their own. Other cells form multicellular organisms — organisms with two or more cells. In **multicellular organisms**, groups of cells often perform specialized functions.

TISSUES

A **tissue** is a group of similar cells that work together. In the human body, for example, **muscle tissue** is made up of similar muscle cells, which act together to relax or contract the muscle.

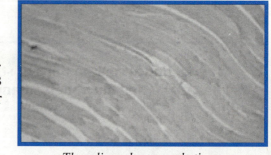

These lines show muscle tissue greatly magnified.

Other types of tissues in the human body include **nervous tissue**, **connective tissue** and **skin tissue**.

ORGANS

An **organ** is made up of two or more tissues working together. Each organ in the human body performs a highly specialized function. For example, your **heart** is one of the most important organs in your body. This fist-sized muscular organ has all four types of tissues. The heart's structure helps it to carry out a special function. Its contractions pump blood throughout your body. Every day, your heart beats about 100,000 times to pump your blood.

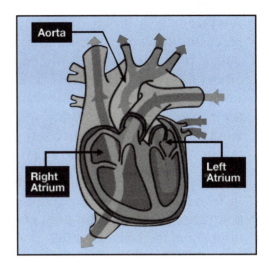

ORGAN SYSTEMS

An **organ system** is a group of organs that work together in an organism to perform a specific function. For example, the heart works together with arteries, veins, and capillaries to carry blood throughout the body. Together, the heart and these blood vessels make up the circulatory system.

Your circulatory system delivers a continuous flow of blood to your body, supplying its cells with oxygen and nutrients, and carrying away wastes.

★ **Arteries** carry oxygen-rich blood from the heart to the rest of the body.
★ **Veins** return blood to the heart.
★ The **capillaries** are very small blood vessels between the arteries and veins. They allow the blood to exchange oxygen with neighboring cells for carbon dioxide. They also enable the blood to deliver nutrients.

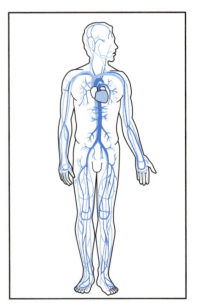

APPLYING WHAT YOU HAVE LEARNED

✦ Create your own glossary. Make a list of each term that follows. Then define each term.

Organism	Organs
Unicellular organism	Heart
Tissues	Organ systems
Muscle tissues	Arteries

THE PROPERTIES OF A SYSTEM

A **system** is a group of structures, cycles or processes that work together. A system always has unique properties that are different from the properties of its parts. The system can accomplish functions that the individual parts cannot achieve on their own. Because a system relies on its parts working together, a system often will not work if one or more of its parts fail or perform poorly.

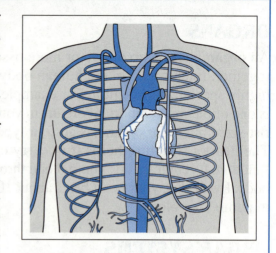

For example, the heart is a single organ with its own characteristics. It is a muscular pump that contracts. It is powerful enough to push blood throughout the body. By itself, however, the heart cannot deliver blood to the rest of the body. It relies on the blood vessels. The circulatory system consists of both these parts: the heart and blood vessels. It is the **system** as a whole that delivers blood throughout the body.

APPLYING WHAT YOU HAVE LEARNED

✦ Create a flow chart showing the path that blood takes through the body. Be sure your flow chart shows the heart, arteries, veins, and capillaries.

✦ How do the properties of a system differ from those of its parts?

ORGAN SYSTEMS OF THE HUMAN BODY

Our bodies are miraculous pieces of machinery. Each of us has ten major organ systems that enable us to live, grow, and reproduce. The structures of these systems are designed to fill special functions. Each system works as a whole to meet our basic needs. These systems are:

— 1 — Skeletal System
— 2 — Muscular System
— 3 — Digestive System
— 4 — Respiratory System
— 5 — Circulatory System
— 6 — Nervous System
— 7 — Excretory System
— 8 — Endocrine System
— 9 — Integumentary System
— 10 — Reproductive System

THE SKELETAL SYSTEM

Our skeletal system consists of bones that provide the framework to support the body. The legs, pelvic bones, spine, shoulder bones, and neck bones support the body's weight. Bones protect our organs from injury. For example, the spine and skull protect the body's central nervous system. The ribs protect the lungs, heart, liver and stomach. Bones permit us to move around. Bones provide a structure for muscles to work against. New blood cells are created inside our bones. Bones also store calcium and other minerals used by our body.

THE MUSCULAR SYSTEM

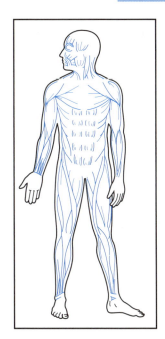

Our muscular system consists of more than 600 muscles that can contract or relax. All muscle action is controlled by your brain, which sends and receives signals through your nervous system. Almost half of the human body's weight is made up of muscle. Skeletal muscles move our bones, while smooth muscles move our organs. Skeletal muscles are attached to bones by stretchy tissue and work in pairs. When one of these muscles contracts, the opposite muscle relaxes, causing the bone to move. The heart is also a powerful muscle — its contractions push blood around the body.

THE DIGESTIVE SYSTEM

Our digestive system consists of organs and glands that allow us to absorb nutrients. Think of the digestive system as a long, twisting tube inside the body. We chew and swallow food, which moves down the esophagus, into the stomach, into the small intestine and finally into large intestine. The mouth and stomach break up and dissolve the food. The liver and other organs provide enzymes to help in this process. In the small intestine, nutrients from digested food are absorbed by capillaries (*tiny blood vessels*). The circulatory system then carries these energy-rich organic molecules to cells throughout the body.

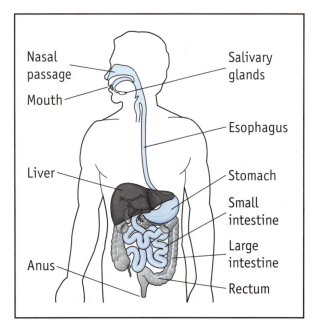

THE RESPIRATORY SYSTEM

Our respiratory system consists of the pharynx, larynx, trachea, bronchi, lungs, and the muscle below the lungs known as the diaphragm. These organs allow us to breathe in air, absorb oxygen, and eliminate carbon dioxide. When you breathe, the diaphram contracts to pull down the lungs. The lungs expand to suck in air. The air reaches tiny sacs in the lungs, which are surrounded by capillaries. Here, oxygen is obtained by the blood, in exchange for carbon dioxide. When you exhale, you expel the carbon dioxide and other gases that your body does not need.

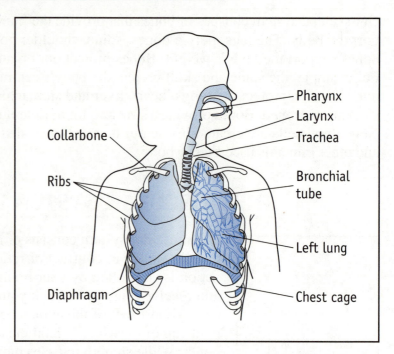

THE CIRCULATORY SYSTEM

As you learned earlier in this chapter, the circulatory system consists of the heart and blood vessels, which deliver oxygen and nutrients throughout the body and carry away wastes. To achieve these functions, the circulatory system works closely with other organ systems. White blood cells in the blood also serve as part of the immune system, which fights invading microbes and cancers.

THE NERVOUS SYSTEM

Our nervous system, the most complex and delicate of all body systems, consists of the brain, sensory organs, spinal cord, and nerves. Electrical impulses run through this system. The nervous system serves as the body's storage and control center. It allows us to think, see, feel, and remember. The nervous system's general function is to collect information about the external environment, analyze the information, and react. From the spinal cord, millions of smaller bunches of nerves branch out to every part of the human body. They give us the sense of touch throughout our body and enable our brain to send messages to our muscles, telling them when to move.

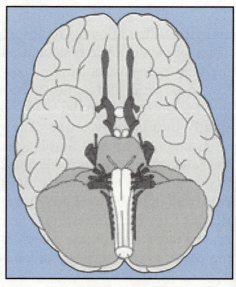

The human brain

THE EXCRETORY SYSTEM

Our excretory system consists of the kidneys and bladder. The excretory system can be thought of as the body's "garbage collector." Your body makes chemical waste products it cannot use. These wastes enter your bloodstream and pass through the kidneys, which act as a filter. Wastes and excess water collected by the kidneys are sent to the bladder. These wastes leave the body as urine. Other parts of the body also work to excrete wastes. The lungs excrete carbon dioxide, while the skin excretes unwanted salts in the form of sweat.

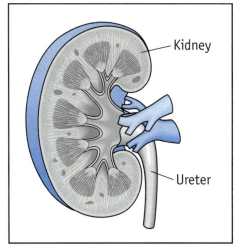

The Excretory Stystem

THE ENDOCRINE SYSTEM

Our endocrine system consists of a series of glands, such as the thyroid gland, the pituitary gland, the pancreas, and the reproductive glands. These glands produce **hormones** — the body's chemical messengers. The endocrine system is responsible for regulating such processes of the human body as growth, the amount of fluids in our bodies, our blood sugar, how energetic we feel, and how we react to stress. Endocrine glands often help in other bodily functions. For example, the pancreas sends fluid to the small intestine to aid digestion.

INTEGUMENTARY SYSTEM

The integumentary system contains the largest organ of the human body, our skin. It is also made up of hair and fingernails. These organs perform an

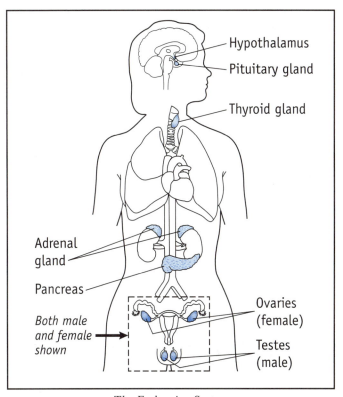

The Endocrine System

important function in our body. They act as barriers against infection and injury, and protect against ultraviolet radiation from the Sun. The skin helps to regulate the body's temperature by perspiring when the body is overheated, and the skin also prevents the body from drying out. Finally, the skin's perspiration helps to remove waste products from the body.

THE REPRODUCTIVE SYSTEM

The human reproductive system differs for males and females. Females have ovaries, where they produce eggs. One egg moves down the fallopian tube each month. Males have testes, where they produce sperm. Sperm is propelled from the male body in fluid, and transferred to the female during sexual intercourse. If one of the sperm fertilizes the egg, the egg will move into the uterus where a baby forms and grows. The baby inherits characteristics from both parents.

APPLYING WHAT YOU HAVE LEARNED

✦ To meet all its needs, your body depends on ten complex systems. Complete the chart summarizing each of the body's main systems:

System	Major Structures	Functions
1. Skeletal		
2. Muscular		
3. Digestive		
4. Respiratory		
5. Circulatory		
6. Nervous		
7. Excretory		
8. Endocrine		
9. Integumentary		
10. Reproductive		

HOW HUMAN ORGAN SYSTEMS INTERACT

In the human body, different organ systems often interact. For example, nerves in the stomach tell the brain when the stomach is empty, making us feel hungry. This interaction of the nervous and digestive systems causes us to eat.

The skeletal, muscular and nervous systems also work together. If we want to move an arm, our brain sends an electric impulse through our nervous system to the nerves in our arm. This causes one pair of muscles in the arm to contract and the opposite pair to relax. Because the skeletal muscles in the arm are attached to the bones in the arm, these muscular contractions cause the bones of the arm to move.

The respiratory and circulatory systems likewise work closely together. The respiratory system brings oxygen into the body and expels carbon dioxide. The lung has tiny sacs that fill with air. These sacs are surrounded by tiny blood vessels. The circulatory system picks up oxygen from the lungs and carries it to the rest of the body. The right side of the heart actually pumps blood directly to the lungs. After leaving the lungs, this oxygen-rich blood returns to the left side of the heart, where it is pumped to the rest of the body.

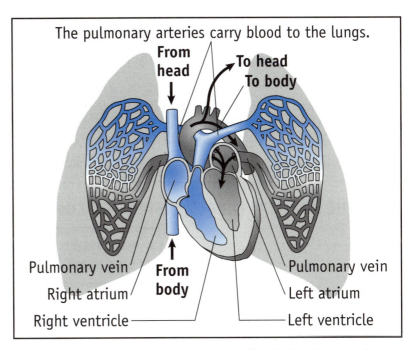

This again shows the close interaction of two different organ systems. Similarly, the circulatory system works closely with the digestive system. Capillaries in the small intestines absorb nutrients, which are sent through the blood to every cell in the body.

APPLYING WHAT YOU HAVE LEARNED

✦ The respiratory and circulatory systems work together to deliver oxygen to cells throughout the body. Provide three other examples of different organ systems in the human body that act together to perform a special function.

First Organ System	Second Organ System	How They Work Together
1.		
2.		
3.		

FEEDBACK MECHANISMS

One way that the human body maintains its equilibrium, or stable internal condition, is through feedback mechanisms. A **feedback mechanism** occurs when the body senses the results of its own actions and then adjusts what it is doing. This allows the body to maintain stable conditions. Feedback mechanisms in the human body often depend on the interaction of two or more organ systems working together to maintain equilibrium.

Controlling Blood Sugar Levels. The endocrine system works closely with the circulatory system to control the level of sugar (*glucose*) in the blood. When the level of glucose in the blood is high, the pancreas automatically produces *more* of the hormone **insulin**.

Insulin tells the liver to process glucose in the blood, turning it into glycogen. More insulin thus lowers the amount of glucose. When the level of glucose in the blood drops, the pancreas automatically produces *less* insulin. The liver then converts less glucose into glycogen, so that the level of glucose in the blood rises again. As a result of these interactions, the level of glucose in the blood stays within a narrow range, keeping the body stable.

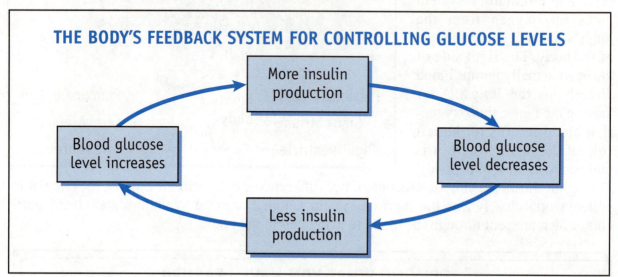

Regulating the Body's Temperature. Another feedback mechanism is the control of your body's internal temperature. The human body must stay within a few degrees of its normal temperature range of 98.6° F. If the body becomes too warm, it begins to perspire. The evaporation of this sweat acts to cool the body. When the temperature of the body cools down, the body reduces the amount of perspiration.

Sweating cools the body to prevent overheating.

Controlling the Pace of the Body's Reactions. The thyroid, another endocrine gland, controls the pace of the body's chemical reactions. When the brain detects that the concentration of thyroid hormone in the blood is too low, it stimulates the thyroid to produce more, increasing the speed of the body's chemical reactions. If these reactions become too fast, the brain instructs the thyroid to make less thyroid hormone, slowing down the speed of the body's chemical reactions.

CHAPTER 10: ORGANISMS AND ORGAN SYSTEMS

APPLYING WHAT YOU HAVE LEARNED

- What is a feedback mechanism?
- Identify another example of a feedback mechanism at work in the body.

WHAT YOU SHOULD KNOW

- ★ You should know that a **tissue** is a group of similar cells that work together.
- ★ You should know that an **organ** is a group of two or more tissues that work together to perform a specific function. The structure of the organ helps it to perform its function.
- ★ You should know that a **system** has its own independent properties that are different from those of its parts.
- ★ You should know that an **organ system** is a group of organs working together to perform a function in an organism. The human body has ten key organ systems that enable it to survive: the skeletal, muscular, digestive, respiratory, circulatory, nervous, excretory, endocrine, integumentary, and reproductive systems.
- ★ You should know that in the body, organ systems often interact and work together.
- ★ You should know that feedback mechanisms help the human body to maintain its **equilibrium** (*a stable balance*).

CHAPTER STUDY CARDS

Important Terminology
- ★ **Tissues.** A group of similar cells with specialized functions.
- ★ **Organs.** Made up of two or more tissues. Each organ performs a highly specialized function.
- ★ **Organ System.** A group of organs that work together in an organism to perform a specific function.
- ★ **Organism.** Any living thing capable of living on its own.
- ★ **Feedback Mechanism.** Occurs when the body senses the results of its actions and adjusts what it is doing, allowing the body to maintain stable conditions.

Interactions of Systems in Humans
Two or more systems often act together to perform a function.
- ★ **Muscular, Skeletal, and Nervous System.** Interact to move an arm.
- ★ **Respiratory and Circulatory System.** Interact to take in oxygen and deliver it to cells throughout the body.
- ★ **Digestive and Circulatory System.** Interact to digest food and absorb and distribute nutrients.
- ★ **Feedback Mechanism.** The endocrine system interacts with other systems to maintain the body's equilibrium, such as the level of glucose in blood.

Organ Systems of the Human Body

In the human body, ten major organ systems each work to meet a specific need:

★ **Skeletal System.** The bones act as a frame to hold our bodies together.
★ **Muscular System.** Made up of skeletal and smooth muscles; moves our bodies.
★ **Digestive System.** Made up of organs and glands that aid the body in digesting food and absorbing nutrients.
★ **Respiratory System.** Made up of lungs and other organs that allow us to obtain oxygen.
★ **Circulatory System.** The heart, veins, arteries and capillaries circulate blood through the body.

Organ Systems of the Human Body

★ **Nervous System.** The brain, sensory organs and nerves allow us to think, see, feel, and react to the environment around us.
★ **Endocrine System.** Produces hormones regulating growth, fluids, blood sugar, and energy levels.
★ **Excretory System.** Kidneys and bladder act as the body's "garbage collector" to filter blood and excrete wastes and water.
★ **Integumentary System.** The skin, hair and fingernails protect the body against infection and injury.
★ **Reproductive System.** Organs that allow humans to produce children.

CHECKING YOUR UNDERSTANDING

1 Some characteristics are listed below:

   1. Includes the stomach, small and large intestine.
   2. Breaks down food into organic molecules.
   3. Absorbs nutrients from digested food.

   Which of these characteristics applies to the human digestive system?

   A   Only 1
   B   Only 2
   C   2 and 3
   D   1, 2, and 3

   ♦ Examine the Question
   ♦ Recall What You Know
   ♦ Apply What You Know

   OBJ. 2
   6.10 (C)

This question asks you to examine three characteristics and asks you to determine which of them applies to the human digestive system. You should recognize that all three of these characteristics — numbers **1**, **2** and **3** — help describe the human digestive system. Thus, choice **D** is the correct answer.

Now try answering some additional questions about organs, organ systems, and the human body.

2 Which is the correct order of organization from smallest to largest?

F cell ⟶ organ ⟶ tissue ⟶ organ system
G tissue ⟶ cell ⟶ organ system ⟶ organ
H tissue ⟶ cell ⟶ organ ⟶ organ system
J cell ⟶ tissue ⟶ organ ⟶ organ system

OBJ. 2
6.10 (C)

3 The chart below illustrates a feedback mechanism in the human body. Which of the following would best complete the blank line?

Regulation of Glucose Levels in the Blood
• Glucose level increases ⟶ Pancreas produces more insulin
• Increased level of insulin ⟶ Glucose level decreases
• Glucose level decreases ⟶ _____

A Pancreas produces more insulin
B Pancreas produces less insulin
C Insulin remains unchanged
D Glucose level rises

OBJ. 2
8.6 (B)

4 Which organ system is responsible for transporting hormones from the endocrine glands to the arms and legs?

F Circulatory
G Excretory
H Digestive
J Nervous

OBJ. 2
8.6 (A)

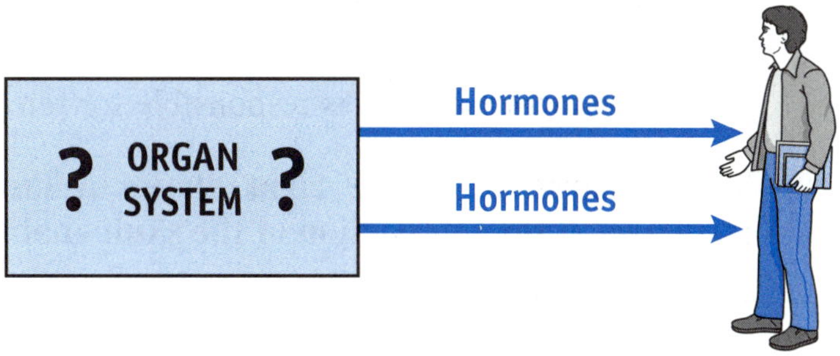

5 Which two systems of the human body interact to send oxygen throughout the body?

A Integumentary and muscular systems
B Digestive and nervous systems
C Respiratory and circulatory systems
D Skeletal and reproductive systems

♦ Examine the Question
♦ Recall What You Know
♦ Apply What You Know

OBJ. 2
8.6 (A)

6 Which organ system of the human body is responsible for secreting hormones that influence bone development and growth?

F Muscular system
G Reproductive system
H Digestive system
J Endocrine system

OBJ. 2
8.6 (A)

7 Which of the following is a direct effect of the interaction of the skeletal, muscular, and nervous systems?

A   Feeling tired after eating a meal
B   Movement of a person's arm
C   Healing of a cut on the foot
D   Increased hormone production

♦ Examine the Question
♦ Recall What You Know
♦ Apply What You Know

OBJ. 2
8.6 (A)

8 A woman is suddenly awakened by the smell of smoke. She gets out of bed and sees smoke and fire coming from the next room. She flees from the burning building. This situation illustrates the interaction of the —

F   digestive and circulatory systems
G   respiratory and circulatory systems
H   endocrine and integumentary systems
J   nervous and muscular systems

OBJ. 2
8.6 (A)

9 To send urine out of the body, muscle contractions in the bladder squeeze the urine through the urethra. This shows that —

A   the body must struggle to rid itself of waste materials
B   organ systems of the body frequently interact
C   all human organs are composed of cells
D   the endocrine system is responsible for removing body wastes

OBJ. 2
6.10 (C)

10 The illustration to the right shows a human skull. The structure at the top portion of the skull enables it to —

F   produce new blood cells
G   allow the movement of joints
H   store minerals like calcium
J   protect the brain from injury

OBJ. 2
6.10 (C)

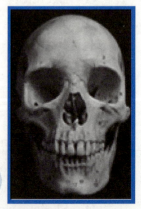

11 Which of the following diagrams best shows the interaction of two organ systems in the human body?

OBJ. 2
8.6 (A)

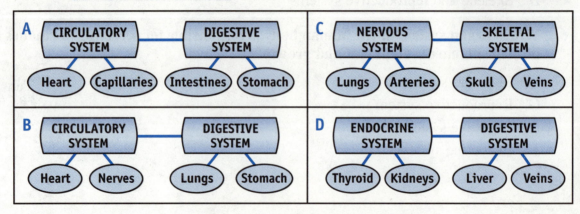

CHAPTER 11

HEREDITY AND ADAPTATION

In this chapter, you will learn how some characteristics of living things are inherited. You will also learn how scientists predict which particular characteristics will be inherited, and how changes in the environment can affect the survival of inherited characteristics.

MAJOR IDEAS

A. Some characteristics, or **traits**, of organisms develop in response to the environment. Other traits are inherited.

B. Inherited traits are determined by **genes**. In sexual reproduction, the offspring receives one gene from each parent for every inherited trait.

C. A **dominant trait** appears in the organism if *either* gene governing that trait is dominant. A **recessive trait** appears in the organism only if *both* genes are recessive.

D. Scientists can predict the possible outcomes of various genetic combinations.

E. DNA molecules provide the hereditary information in each gene. DNA molecules sometimes change or mutate. Such changes allow for genetic differences.

F. Environmental changes can affect the survival of individual organisms. Organisms with favorable characteristics are more likely to survive environmental changes. The proportion of individuals in a population with favorable hereditary characteristics gradually increases over time. As a result, a species slowly changes in response to environmental change.

THE MECHANICS OF HEREDITY

A **trait** is a characteristic of a living organism. For example, you may be tall, have brown eyes, and like pancakes for breakfast. These are all examples of traits.

LEARNED AND INHERITED TRAITS

Some traits develop in response to the environment. For example, you may have learned to like eating pancakes with maple syrup. Learned traits sometimes change. For example, you may get sick of eating pancakes for breakfast every morning. Other traits are **inherited**. They are passed on from the parents to their children. Your height was inherited from your parents. So, too, was the color of your eyes. No matter how hard you try, you cannot change your height or the color of your eyes.

APPLYING WHAT YOU HAVE LEARNED

◆ Some people show dimples in their cheeks when they smile. When people put their hands together and interlock their fingers, they place either their left or right thumb on top. Other people can curl the sides of their tongue. These are all **inherited traits**. A person who cannot curl his or her tongue cannot learn to do so. Pick a classmate and take an inventory of these easily observable traits. Compare your traits with those of your partner.

	Your Traits	Classmate's Traits
1. Dimples when smiling	☐	☐
2. Right thumb on top	☐	☐
3. Can curl tongue	☐	☐

HOW TRAITS ARE INHERITED

We have **inherited traits** because of our genes. A **gene** is the part of a chromosome that determines a specific trait. Each inherited trait is controlled by two genes. You inherit one of those genes from your mother and one from your father.

An Austrian monk, **Gregor Mendel**, discovered the basic laws of heredity in the 1850s. Mendel was interested in how parents pass on their traits to their offspring. Mendel conducted experiments with pea plants. He chose pea plants to study for several reasons. Pea plants were small, easy to grow, produced large numbers of offspring and matured quickly.

Gregor Mendel

Mendel found that pea plants often have contrasting traits. They can be tall or short. Their flowers can be white or purple. Their seeds can be smooth or wrinkled. Mendel found that pea plants reproduce when pollen (*male*) is rubbed onto the plant's flowers (*female*). A pea plant can pollinate itself or can pollinate another plant.

Mendel rubbed pollen from each plant onto its own flowers and did this for several generations. In this way, he obtained plants that he knew were pure for each trait. Then he mixed his plants by rubbing pollen from one plant onto another. He made some surprising discoveries.

- ★ **Dominant Traits.** For each pair of contrasting traits, one trait was **dominant**. For example, if he crossed a pea plant having purple flowers with a plant having white flowers, the offspring always had purple flowers. If he crossed a pea plant with smooth seeds with one with wrinkled seeds, the offspring always had smooth seeds. The plant would show the dominant trait if it inherited that trait from *either* parent.

- ★ **Recessive Traits** He also found the other trait was **recessive**. When a purple flower was crossed with a white flower, Mendel found their offspring always had purple flowers. However, if he crossed their offspring, the recessive trait reappeared: one quarter of the third generation would have white flowers.

PREDICTING POSSIBLE OUTCOMES
OF GENETIC COMBINATIONS

From Mendel's discoveries and later research, scientists are now able to make predictions about what traits will be inherited by an organism. If the organism has sexual reproduction, then each trait is governed by two genes. One gene is inherited from each parent. The recessive trait only appears in the organism if the recessive gene is inherited from both parents. If the organism inherits the dominant gene from either parent, it "masks" or hides the recessive trait.

Scientists use a diagram, called a **Punnett square**, to predict which genes will be inherited. For each gene, a **capital letter** is used to indicate the dominant trait. A **lower case** letter is used to show the recessive trait.

Suppose a scientist decides to cross a pea plant with purple flowers and a pea plant with white flowers. Each plant comes from several generations of plants with those same flower colors. Therefore, the scientist knows that each plant has two of the same type of gene. The purple-flowered plant has two purple-flower genes, which are represented by two capital letters: **PP**.

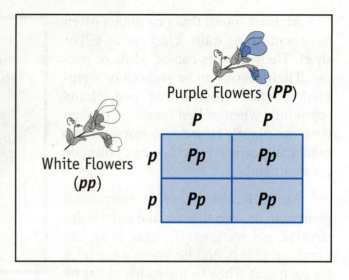

The white-flowered plant has two white-flower genes, represented by two lower case letters: **pp**. The Punnett square above shows what genetic combinations the cross between these two plants can have.

In this case, even though one of the parents has white flowers, all of the possible offspring will have purple flowers — *Pp*. This is because having purple flowers is a dominant trait, and each offspring will inherit one purple-flower gene (*P*).

Now, look at what happens when the offspring of this combination pollinate each other:

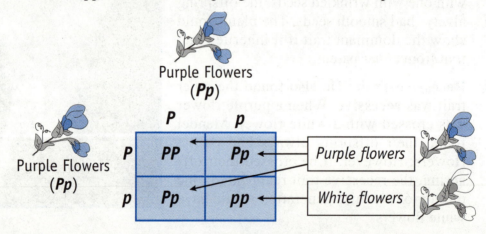

In this case, both parents have purple flowers. However, each one also carries the **recessive gene** (*white flowers*). From their cross-pollination, four possible combinations can result:

★ In one combination, the offspring inherits the dominant trait (*P*) from both parents. This offspring (*PP*) will have purple flowers.

★ In two of the combinations, the offspring would inherit the dominant gene (*P*) from one parent and the recessive gene (*p*) from the other. The offspring from this combination (*Pp*) will also have purple flowers.

★ In the final combination, the offspring inherits the recessive gene (*p*) from each parent. In this case, the offspring will have white flowers (*pp*). This situation will occur in about one-fourth of the offspring of these two parents.

A Punnett square is often used to predict the probability of inheriting a particular trait. In this case, one-quarter of the offspring will be (**PP**), one half of the offspring will be (**Pp**), and one-quarter of the offspring will be (**pp**).

APPLYING WHAT YOU HAVE LEARNED

✦ For a certain type of plant, tallness (**T**) is the dominant gene. Complete the Punnett square below for a tall plant (**TT**) crossed with a short plant (**tt**).

Tall Plant (**TT**)

Short Plant (**tt**)

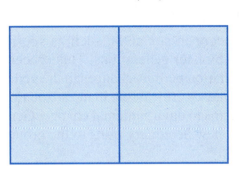

✦ What proportion of their offspring will be tall plants? Explain your answer.

THE ROLE OF DNA IN HEREDITY

Although Mendel discovered some of the basic laws of **heredity**, he did not know why they worked. A century later, two scientists discovered the molecular basis of heredity. In 1953, **Francis Crick**, a graduate student at Cambridge University, and **Dr. James Watson**, a biochemist, developed the first model for the structure of the DNA molecule. Watson and Crick relied in part on the work of **Rosalind Franklin**, who had taken x-ray photographs of DNA.

Watson and Crick concluded that all living cells contain **DNA** (*deoxyribonucleic acid*). The DNA of every living organism, from the simplest bacteria to human beings, is made up of these same chemical ingredients. Crick and Watson found that this molecule has a unique ability to copy itself during cell division. It also encodes all genetic information, providing instructions for the operation of every cell in a living organism. The DNA molecule resembles a "double helix," much like a twisted ladder.

Francis Crick and James Watson (1953)

GENETIC VARIATION AND MUTATION

When sex cells divide, their chromosomes twist around and actually swap some genes. In addition, DNA can sometimes be damaged by ultraviolet radiation, by environmental factors, or by a random error that occurs while copying itself. Such a change in DNA is known as a **mutation**.

Some mutations can cause serious disorders, like cancer. Cancer is caused by mutations that lead to unchecked cell divisions. Cancer cells keep dividing continuously, causing tissue to grow into a tumor.

When mutations occur in sex cells, such as sperm or eggs, they are inherited by later generations. This process of genetic variation and mutation permits changes in the genes of organisms and actually plays a critical role in the adaptation of groups of related organisms to environmental change. Genetic mutations can introduce new hereditary traits to the group.

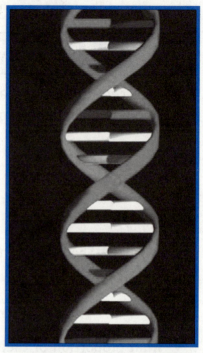

A model of the DNA molecule

APPLYING WHAT YOU HAVE LEARNED

✦ What role does DNA play in heredity?

✦ What are some of the factors that can cause DNA to become "damaged"?

ADAPTATION TO ENVIRONMENTAL CHANGE

Over time, Earth's environment changes. For example, a region may become cooler or drier. Living organisms must adapt to these changes in order to survive. One way that an individual organism can adapt is to learn special behaviors to cope with its environment.

Species and Populations. All of the organisms of the same type who are able to have offspring together form a **species**. All of the organisms of the same species living in an area are known as a **population**. Like individuals, large populations of organisms must adapt to environmental change if they are to survive. Heredity plays an important role in this process.

Natural Selection. The process known as **natural selection** helps explain how species adapt to environmental change. For example, some rabbits naturally have thick fur, while others have thin fur. This is a random difference that may have first occurred as a result of genetic mutation.

If the climate of a region changes and becomes colder, the rabbits with thick fur will have a better chance of surviving than the rabbits with thin fur. The rabbits with thicker fur will better withstand cold temperatures and have longer periods outside their homes to find food. More of the thick-furred rabbits will survive and reproduce than the thin-furred rabbits. Gradually, they will make up a larger proportion of the rabbit population in that area.

Random genetic differences thus become important when they give some individuals inherited traits that are useful in adapting to a particular environment. Organisms with those traits are more likely to survive and reproduce, passing their favorable genetic traits on to their offspring. Eventually, all of the surviving offspring in the species will have those traits.

Scientists refer to the view that a species adapts to its environment through random genetic change and natural selection as the **theory of evolution**. Many scientists believe that this process helps to explain the great diversity of life found on our planet.

Why is thick fur on a rabbit beneficial in colder climates?

Extinction. Species can adapt to environmental change through genetic mutation and natural selection. This process generally takes millions of years. Sometimes, however, natural selection may fail to work. The environment may change so quickly that a species is unable to adapt. For example, many scientists believe that about 65 million years ago, a giant meteor crashed into Earth. Dust from the impact covered Earth's atmosphere, blocking much of the sunlight. Plants died and, after living on Earth for about 165 million years, the dinosaurs that fed on these plants also died out. When a species fails to adapt and dies out, scientists say it has become **extinct**. A species that is close to extinction is known as an **endangered** species. Today, many species are endangered by the effects of human activities.

Scientists believe dinosaurs may have become extinct because they failed to adapt to changing conditions.

APPLYING WHAT YOU HAVE LEARNED

In the 1830s, a young naturalist, **Charles Darwin**, traveled to the Galapagos Islands off the coast of South America. Darwin was struck by the wide variety of finches, a type of bird, with different beaks. Darwin found that some had narrow beaks that could fit into the gaps between stones to find insects. Others had large beaks useful for cracking shells and nuts found on the island. Each type of beak appeared suited to finding and using a different food source. Darwin reasoned these birds all came from a common ancestor, which had flown to the Galapagos Islands from South America.

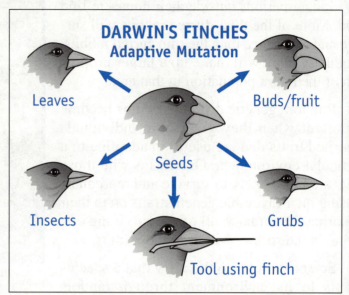

◆ How might genetic mutation and natural selection help explain these differences if all these birds came from the same common ancestor?

WHAT YOU SHOULD KNOW

★ You should know that some **traits** are inherited.

★ You should know that the offspring of sexual reproduction inherit one gene from each parent for each trait. Together, the two genes will determine the trait of the offspring. If the offspring inherits the **dominant gene** from either parent, the **dominant trait** will appear. If the offspring inherits the **recessive gene** from both parents, the **recessive trait** will appear.

★ You should know that scientists use a **Punnett square** to predict the possible outcomes of genetic combinations.

★ You should know that **DNA** molecules provide the hereditary information in each gene.

★ You should know that organisms with favorable hereditary traits are more likely to survive and pass their traits on to their offspring. This process of **natural selection** helps species slowly adapt to environmental change.

CHAPTER STUDY CARDS

Heredity

- ★ **Inherited Trait.** A trait an organism inherits from its parents, such as height or eye color.
- ★ **Gene.** The part of a chromosome that governs a particular trait.
- ★ **Punnett Square.** Diagram used to predict outcomes of genetic combinations.
- ★ **Dominant Trait.** Appears if it inherits the gene for that trait from either parent; shown by capital letters on a Punnett square.
- ★ **Recessive Trait.** Appears only if it inherits that trait from both parents; shown by lower case letters on a Punnett Square.

Adaptation

- ★ **Genetic Mutation.** Change in gene caused by environmental damage or random error.
- ★ **Environmental Change.** When the environment changes, such as the climate becoming colder or dryer.
- ★ **Natural Selection.** Organisms with favorable hereditary traits are more likely to survive and reproduce than other organisms; these organisms gradually increase their proportion of a species.
- ★ **Species.** Group of similar organisms that can have children together.
- ★ **Population.** All the members of a species living in an area.

CHECKING YOUR UNDERSTANDING

1 Most pea plants have green coverings on their seeds, known as pods. The dominant gene for pod color is green (G). A pea plant will have a yellow pod only if it has two recessive genes (gg).

Which of the following Punnett squares correctly predicts the offspring of a cross between two pea plants with green pods?

OBJ. 2
8.11 (C)

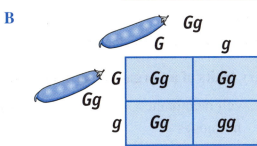

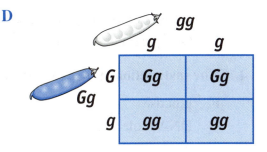

HINT This question tests your understanding of heredity. You must interpret information in a Punnett square. Choice D is incorrect because it uses (*gg*) for one of the parents. Choice B is incorrect because it shows (*Gg*) for offspring inheriting *G* genes from each parent. Which remaining choice is correct?

Now try answering some additional questions dealing with heredity and adaptation.

2. In *Serinus canaria* (canaries), the gene for singing (S) is dominant over the gene for non-singing (s).

If the two canaries in the picture were crossed, what percent of their offspring would be expected to sing? Record and bubble your answer on the grid below.

- Examine the Question
- Recall What You Know
- Apply What You Know

OBJ. 2
8.11 (C)

3. For a certain type of cat, the gene for black fur (B) is dominant while the gene for brown fur (b) is recessive. Two cats with black fur produce an offspring with brown fur. Which best describes the genes of the parent cats?

  A One of the parents carries the recessive gene (b), but the other parent does not.
  B Neither parent carries the recessive gene (b).
  C Both parents carry one recessive gene (b).
  D One parent carries two recessive genes (bb), but the other does not.

OBJ. 2
8.11 (C)

4. The instructions for the various hereditary traits of an organism are found in its —

  F glucose
  G DNA molecules
  H cell membrane
  J hormones

OBJ. 2
8.10 (C)

5 A large population of houseflies was sprayed with a newly developed, fast-acting insecticide. The appearance of increasing numbers of houseflies that are unaffected by this insecticide supports the view that —

- A a species' traits tend to remain unchanged
- B the environment for houseflies does not change
- C genetic variation permits species adaptation
- D houseflies are a unique form of bacteria

6 Which of the following is the best example of an inherited trait?

- F Mrs. Smith is five feet tall.
- G Aiesha has learned about algebra.
- H Dwayne eats ice cream every night.
- J Mr. Smith is worried about his daughter's grades.

7 For pea plants, having purple flowers (*P*) is a dominant trait. A pea plant with two dominant genes (*PP*) is crossed with a pea plant with one dominant and one recessive gene (*Pp*). What proportion of their offspring will have purple flowers?

- A 0%
- B 25%
- C 75%
- D 100%

8 In a species of plant, the sudden appearance of one plant with a different leaf color would most likely be the result of —

- F photosynthesis
- G genetic mutation
- H slow environmental change
- J asexual reproduction

9 Some rabbits have white fur while other rabbits in the same area have brown fur. If the climate of the area changes so that there is increased snow in winter, which of the following is most likely to happen?

- A More brown rabbits will survive than white rabbits because they can find each other more quickly.
- B More white rabbits will survive than brown rabbits because they are better able to withstand the cold temperatures.
- C More white rabbits will survive than brown rabbits because animals that hunt rabbits cannot easily see white rabbits against the snow.
- D The proportion of white rabbits and brown rabbits will stay about the same.

10 Dolphins communicate by making sounds in the water. Dolphins that cannot make sounds are less likely to survive or have children. As a result, over time the number of dolphins that cannot make sounds is gradually reduced. This example illustrates the process of —

- F cellular respiration
- G genetic mutation
- H natural selection
- J environmental change

CHAPTER 12

ECOSYSTEMS

All organisms depend on both the environment and other organisms to survive. **Ecology** is the study of relationships between living organisms and their environment. Ecologists often study how different organisms live and interact in a specific area. All of the living organisms in an area, together with the nonliving environment, make up an *ecological system*, often known as an **ecosystem**.

Guadalupe Mountains National Park in Texas forms an ecosystem with a vast diversity of life.

The size of an ecosystem can vary greatly, from a small pond to a vast forest. Different ecosystems are often separated by geographical barriers such as deserts, mountains or oceans. However, the borders of ecosystems are not rigid, and one ecosystem often blends into another. An ecosystem such as a tropical rainforest may also have several smaller ecosystems within it, such as a forest canopy ecosystem and a forest floor ecosystem.

MAJOR IDEAS

A. An **ecosystem** is a community of living organisms in an area and their environment.

B. Different environments support different types of organisms.

C. The organisms of an ecosystem are **interdependent**.

D. Ecosystems recycle both matter and energy. **Producers**, **consumers**, and **decomposers** live together and use existing resources in an ecosystem.

E. Drastic events, such as a fire or climate change, may trigger a **succession** of changes to an ecosystem.

TYPES OF ECOSYSTEMS

There are many types of ecosystems. Some ecosystems are on land. These include forest, grassland, desert, and tundra ecosystems. Ecosystems in water are known as **aquatic ecosystems**. These include freshwater ecosystems like the Great Lakes, and ocean zone ecosystems like the Continental Shelf, Great Reef Barrier, and the mid-ocean floor.

Alligators play an important role in the ecological balance of the Everglades wetlands.

Community and Population in an Ecosystem. The living organisms of different species found in a single ecosystem are referred to as a **community**. For example, all of the trees, squirrels, foxes, deer and other organisms living in a temperate forest ecosystem make up its community. All of the organisms of the same species in a particular ecosystem are known as the **population** of that species. For example, all of the alligators found in the Everglades wetlands make up that ecosystem's *alligator population*.

APPLYING WHAT YOU HAVE LEARNED

◆ Think of an ecosystem in the area where you live. Identify the ecosystem's location and some of the different organisms that live in that ecosystem.

DIFFERENT ENVIRONMENTS SUPPORT DIFFERENT ECOSYSTEMS

Environmental factors often determine what kind of organisms can live in a particular area. These environmental factors include temperature, the availability of water, humidity, the type of soil and minerals, the amount of sunlight, and the concentration of oxygen and other gases. Differences in these environmental factors give rise to different types of ecosystems. For example, the following five types of ecosystems are found on Earth's land areas:

TEMPERATE FORESTS

Temperate forests develop in mid-latitude regions where there are 30 to 60 inches of rain each year. The four seasons are marked by moderate temperatures and cool winters. Trees change colors in fall and lose their leaves in winter. There is a wide range of plant and animal life. Insects, spiders, slugs, frogs, turtles, and small and large mammals are common. Animals in this ecosystem must be able to adapt to changing seasons. Some animals migrate or hibernate in the winter.

TROPICAL RAIN FORESTS

Tropical rain forests develop in tropical areas near the equator where there is ample rainfall and warm temperatures year-round. Tropical rain forests cover about two percent of the globe. Large trees cover these areas with their leaves, forming a **canopy**. Despite the rapid growth of trees, the topsoil is actually very thin. Tropical rain forests are marked by a great abundance of animal and plant life with greater biological diversity than in any other type of ecosystem. They are home to more than half the world's living plant and animal species, including many unique plants, insects, birds, reptiles and mammals.

DESERTS

Deserts are regions that receive less than 10 inches of rainfall annually. Deserts in the tropical latitudes, such as the Sahara Desert, have their own special forms of plant and animal life. Desert species have adapted to the lack of water and extremes of temperature. A cactus, for example, stores water in its stem. Its spikes prevent thirsty animals from eating it. A camel can go for long periods of time without water. Many other insect and reptile species have also adapted well to desert conditions — for example, many are only active at night.

A desert ecosystem in daylight

A desert ecosystem at night

GRASSLANDS

Grassland areas exist where the climate is drier than a forest but wetter than a desert. There is not enough rainfall to support large numbers of trees. Instead, grasses dominate, as well as large grazing animals like cattle, antelope or bison.

TUNDRAS

Tundras are found closer to polar regions. There the soil is so cold that trees cannot grow. Below the thin layer of tundra soil is a permanent frozen layer of ground. Tundras constitute a distinct type of ecosystem, with their own plant and animal life, including grasses and small shrubs. Every animal in the tundra must adapt in order to survive. Some grow thick fur, while others find a place to hibernate during the cold winters. Large mammals and birds migrate to these regions in the warmer spring and summer months.

Cattle are found mainly in grassland areas.

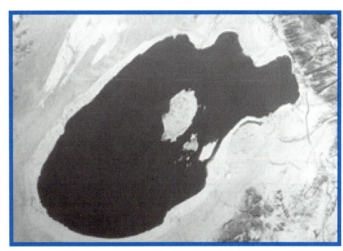
In a tundra, much of the ground is frozen.

APPLYING WHAT YOU HAVE LEARNED

✦ Select **one** of the five ecosystems you have just read about.
  • For your selection, list some of the major plant and animal species found there.
  • Describe some of the unique characteristics these plants and animals possess in order to exist in those environmental conditions.

There is equal diversity among aquatic ecosystems. Oceans, for example, have different life forms at various water depths. Less sunlight reaches deeper layers of the ocean, affecting the forms of life that can survive there. **Estuaries** — places where freshwater rivers flow into the sea — form their own ecosystems; so, too, do freshwater lakes, rivers, and streams.

INTERACTION OF ORGANISMS IN AN ECOSYSTEM

All animals and plants require resources — nutrients, energy, and space — to survive. Many of these resources are provided by other living things. Every ecosystem therefore contains different populations of species that interact together.

Predators. Sometimes, one organism — known as a **predator** — captures and kills another organism, known as its **prey**. Predators like wolves and tigers usually have special characteristics like sharp teeth and the ability to run fast to hunt their prey. Prey also have special physical characteristics, like eyes on the sides of the head, to avoid being caught.

Parasites. A **parasite** feeds on another organism — known as the **host**. Unlike predators, the parasite does not kill the host, at least not immediately. Some parasites attach themselves on the surface of the host, such as ticks. Others, like tapeworms, live inside the host.

Cooperation. Sometimes there is a cooperative relationship in which two or more species mutually benefit. For example, flowering plants provide nectar to bees and other insects. The insects carry pollen from one plant to the flowers of other plants, leading to cross-fertilization. Both the plants and the insects benefit.

Competition. Similar species may compete for the same resources. For example, cows and sheep may eat the same grasses. An increase in one species may hurt the other.

APPLYING WHAT YOU HAVE LEARNED

◆ Provide one example for each of these four types of interactions:

Interaction	Example	Interaction	Example
Predator		Cooperation	
Parasite		Competition	

EQUILIBRIUM AND CHANGE IN ECOSYSTEMS

The existence of limited resources (*water, organic molecules, oxygen, carbon dioxide, space*), competition with other species, and the loss of individuals to predators limits the growth of each species in an ecosystem. There may be periodic changes within the community of an ecosystem, but a delicate **equilibrium** (*balance*) is usually reached among its different species.

This equilibrium can be upset by drastic natural disasters, climate changes, the introduction of new species, or the disappearance of an existing species. Such changes can have effects that ripple through the ecosystem. For example, wolves may keep the number of deer in a forest ecosystem in check. If human hunters kill the wolves, the deer population expands too rapidly, threatening plant life and other plant-eating animals in the ecosystem.

Human activity can have important consequences on an environment. When humans cut down trees to build homes, this may deprive some animals of their natural habitat. Forced to find a new environment to live in, they may not be able to adapt to living elsewhere. As a result, some animals become **endangered species**. Not all human actions have such drastic effects, and there are often varying degrees of consequence.

Ecological Succession. More drastic changes in the environment often lead to a **succession** (*series*) of changes in an ecosystem. Different organisms appear as the environment changes. This process is referred to as **ecological succession**. For example, a fire may destroy ancient trees in a temperate forest ecosystem. Weeds and grasses will quickly spring up in the ashes, which contain valuable nutrients. Next, shrubs and taller plants will grow, blocking the light for the grasses. Finally, pine trees and trees that shed their leaves in fall will take root, replacing the shrubs. A new temperate forest develops, beginning the cycle all over again.

After fire destroys this forest, a succession of changes will occur.

THE FLOW OF RESOURCES THROUGH AN ECOSYSTEM

Energy and matter — water, nutrients, and waste — are continually recycled within an ecosystem. Ecologists are often able to trace the flow of energy and nutrients through an ecosystem.

PRODUCERS

The basic source of energy for all ecosystems is sunlight. **Producers** capture this energy and turn it into carbohydrates through photosynthesis. In land ecosystems, the producers are green plants. These plants use the energy of sunlight in photosynthesis, but they also require carbon dioxide, water, and nutrients from the soil. In aquatic ecosystems, microscopic phytoplankton use photosynthesis to obtain energy.

CONSUMERS

Unlike plants, animals cannot make their own food. They are **consumers** that eat other organisms to obtain organic molecules containing energy.

★ Some animals, known as **herbivores**, eat only plants or phytoplankton.
★ **Carnivores**, such as tigers, only eat other animals.
★ **Omnivores**, such as dogs and humans, eat both plants and animals.

DECOMPOSERS

Some organisms, like vultures, ants, and bacteria, consume dead organisms or animal wastes. Decomposers, such as worms, bacteria and fungi, break down dead organisms and animal wastes into organic molecules. These organic molecules, especially nitrates, are then used by plants.

A **food chain** or **food web** is a diagram tracing the flow of energy and nutrients through a single ecosystem. It shows the specific links between organ-

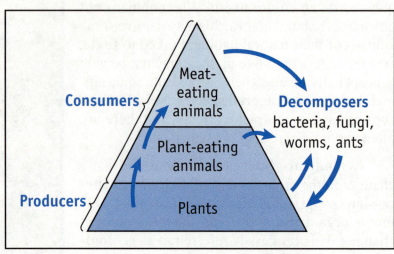

Energy and nutrients are continually recycled in an ecosystem.

isms in the ecosystem. For example, the grasses (*producers*), rabbits (*consumers*), coyotes (*consumers*) and worms, bacteria and fungi (*decomposers*) on a prairie form a single food chain. Energy flows from the sun to the grasses. Rabbits feed on the grasses and absorb their energy and nutrients. Coyotes eat some of the rabbits and absorb their energy and nutrients. When the rabbits and coyotes die, worms, bacteria, and fungi decompose their bodies and enrich the soil for the grasses.

THE FOOD CHAIN IN A PRAIRIE ECOSYSTEM

Water, carbon dioxide, and nitrates are similarly recycled through an ecosystem. For example, water is stored in the ocean, freshwater sources, or underground. It evaporates into the atmosphere, where it becomes water vapor. Later, this water comes back to Earth's surface as rain, snow, sleet or fog. Plants take this water through their roots, and animals drink water. All organisms then release this water back into the environment.

APPLYING WHAT YOU HAVE LEARNED

✦ Draw a food chain or web showing yourself as one of the consumers. In your diagram, show what you ate for dinner last night and consider where the energy in those ingredients came from.

WHAT YOU SHOULD KNOW

★ You should know that an **ecosystem** is a distinct area with its own nonliving environment and a community of living organisms.

★ You should know that different environments support different types of organisms. Five typical land ecosystems based on environmental factors are: temperate forest, tropical rain forest, grassland, desert, and tundra.

★ You should know that the organisms of an ecosystem are **interdependent** — they depend upon each other.

★ You should know that energy and nutrients are cycled and recycled through an ecosystem. **Producers** (*plants*) use nitrates from the soil and sunlight to make organic molecules. **Consumers** (*animals*) eat producers. **Decomposers** (*worms, fungi, bacteria*) break down dead plants and animals into organic molecules.

★ You should know that drastic events, such as fires or climate changes, can trigger the process of **ecological succession** (*a series of changes in an ecosystem*).

CHAPTER STUDY CARDS

Ecosystems
An ecosystem is a community of living organisms and their environment in a specific area.
- ★ **Nonliving Environmental Factors.** Non-living factors influence an ecosystem, such as temperatures, sunlight, and soil.
- ★ **Community.** All the organisms found in a single ecosystem.
- ★ **Population.** All of the organisms of the same species in a particular ecosystem.
- ★ **Examples of Land Ecosystems:** Temperate forest, tropical rain forest, grassland, desert, and tundra.

Flow of Resources in an Ecosystem
- ★ **Interaction of Organisms.** Predators, parasites, competition and cooperation.
- ★ **Recycling of Energy and Nutrients.**
  - **Producers** (*plants*) obtain energy from sunlight, water and nitrates from soil.
  - **Consumers** (*animals*) eat plants or other animals; provide nitrates and CO_2 to be used by plants.
  - **Decomposers** (*bacteria, fungi*) break down dead organisms into organic compounds.
- ★ **Ecological Succession:** Drastic events, like fire, bring a series of changes to an ecosystem.

CHECKING YOUR UNDERSTANDING

1. **Which of the illustrations below best shows an ecosystem?**

- ♦ Examine the Question
- ♦ Recall What You Know
- ♦ Apply What You Know

OBJ. 2
7.11 (C)

This question examines your understanding of an ecosystem. Recall that an ecosystem consists of all of the living organisms in an area, together with their nonliving environment. Although illustration A shows different species, they do not represent an ecosystem of organisms and their nonliving environment. Only illustration D includes both the organisms and the nonliving environment necessary for an ecosystem.

Now try answering some additional questions on your own about ecosystems.

2. **Which of the following can be described as a species population in an ecosystem?**

   F   All the honey bees in an orchard
   G   All the plants and animals in a pond
   H   All nonliving environmental conditions
   J   All living things in the Earth's atmosphere

OBJ. 2
8.6 (C)

| Rain in the Sahel grasslands, south of the Sahara Desert, sharply declines. | → | Grasses in the Sahel die. | → | Sheep and cattle, which depend on the grass, also die. | → | New plants and animals, common in desert conditions, emerge. |

3 The example above best illustrates —

   A equilibrium in an ecosystem
   B the recycling of nutrients in an ecosystem
   C the extinction of a species
   D a succession of ecological changes

OBJ. 2
7.12 (D)

SNOWSHOE HARES

A Snowshoe hare

The snowshoe hare is the prey of many woodland animals. Foxes, coyotes, and weasels hunt the hares by day, while the great horned and barred owls hunt them at night. The snowshoe hare survives partly through the effect of its changing fur color, creating a camouflage. Some favorite foods of the hare during the winter months are twigs of maple, birch, and apple trees. Grasses and clover replace this diet during the spring and summer. Hares tend to feed during the hours of dusk and dawn, when the light is low and their predators are inactive.

4 Based on its diet, the snowshoe hare would probably be classified as a —

   F consumer
   G predator
   H decomposer
   J producer

♦ Examine the Question
♦ Recall What You Know
♦ Apply What You Know

OBJ. 2
8.6 (C)

5 Since foxes, coyotes, weasels and owls hunt snowshoe hares, these animals would most likely be classified as —

   A decomposers
   B parasites
   C predators
   D producers

OBJ. 2
7.12 (B)

Use the information in the following table to answer question 6.

Species of Bear	Location	Food Sources
Brown Bear	Europe, Asia, and North America	fruits, nuts, roots, insects, fish, small animals
Black Bear	North America	fruits, nuts, plant roots, bee honey, insects, rats, mice, fish
Polar Bear	Arctic	seals, fish, seabirds, hares, caribou, musk oxen
Panda Bear	China	bamboo stems and leaves

6 Which of these four species of bear would be most in danger of becoming extinct if one of its food sources became unavailable?

F Brown bear
G Black bear
H Polar bear
J Panda bear

OBJ. 2
8.11 (A)

7 Which evidence best shows that the organisms of an ecosystem are interdependent?

A Plants are able to convert light into chemical energy.
B Different environments support different species of organisms.
C Parasites often kill their hosts.
D Animals must eat plants while plants require nitrates from decomposed animals.

OBJ. 2
7.12 (B)

8 Which of the following would best complete the food web illustrated on the right?

F Trees
G Fish
H Rabbit
J Hawk

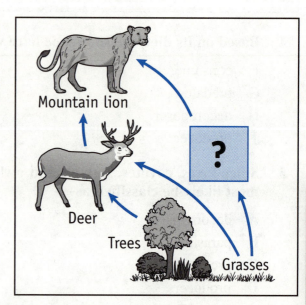

♦ Examine the Question
♦ Recall What You Know
♦ Apply What You Know

OBJ. 2
7.12 (B)

9 **In a grasslands ecosystem, lions are predators that eat antelope. What impact would the sudden disappearance of lions most likely have on this ecosystem?**

   A The grasses would grow larger.
   B The climate would grow increasingly warmer.
   C More antelopes would eat the grasses.
   D Antelopes would start to eat each other.

   OBJ. 2
   8.6 (C)

CHECKLIST OF OBJECTIVES IN THIS UNIT

Place a check mark (✔) next to those objectives you understand. If you have trouble recalling any of these, review the chapter in the brackets. **You should be able to:**

❑ explain how all organisms are composed of cells, which carry on functions to sustain life. **[Chapter 9]**

❑ identify how structure complements function at different levels of organization in living systems, including organs, organ systems, organisms, and populations. **[Chapters 9, 10, 11 and 12]**

❑ identify that radiant energy from the Sun is transferred into chemical energy through the process of photosynthesis. **[Chapter 9]**

❑ describe how the properties of a system are different from the properties of its parts. **[Chapter 10]**

❑ identify feedback mechanisms that maintain the equilibrium of systems, such as body temperature, turgor pressure, and chemical reactions. **[Chapter 10]**

❑ describe interactions among organ systems in the human organism. **[Chapter 10]**

❑ distinguish between inherited traits and other characteristics that result from interactions with the environment. **[Chapter 11]**

❑ make predictions about the possible outcomes of various genetic combinations of inherited characteristics. **[Chapter 11]**

❑ show how change in environmental conditions can affect the survival of individuals and species. **[Chapter 11]**

❑ describe how different environments support different varieties of organisms. **[Chapter 12]**

❑ describe interactions within ecosystems. **[Chapter 12]**

❑ describe how organisms, including producers, consumers, and decomposers, live together in an environment and use existing resources. **[Chapter 12]**

❑ observe and describe the role of ecological succession in ecosystems. **[Chapter 12]**

EARTH AND SPACE SYSTEMS

UNIT 6

In this unit, you will review what you need to know for the **Middle School TAKS in Science** about Earth and space systems. You will study the processes that occur on our planet, and the relationship of our planet to the rest of the universe.

For countless centuries, ancient peoples gazed up at the night sky and wondered what they were seeing. They saw fast-moving points of light, which the ancient Greeks called "planets," meaning wanderers. Today, scientists continue to study the planets, including our own, to understand our place in the universe.

The planet Earth as seen from the moon

★ **Chapter 13: The Universe**
In this chapter, you will learn about the universe, the nature of stars, and the movements of the Earth and moon. You will also learn about the rotation and tilt of Earth and its impact on this planet.

★ **Chapter 14: Planet Earth: Cycles, Systems, and Interactions**
In this chapter, you will learn about Earth's systems and how they interact. You will learn about rock and water cycles, and how the Earth's surface is shaped by tectonic plate movement, weathering, and erosion. You will also learn how the air, ocean, and solar radiation interact to create weather. Finally, you will examine the impact of human activity on Earth's systems.

CHAPTER 13

THE UNIVERSE

Astronomy, the study of the stars, planets, and outer space, is the oldest science. In this chapter, you will learn about stars, galaxies, and our solar system.

MAJOR IDEAS

A. The **universe** consists of all matter, energy and space.

B. **Stars** produce energy through nuclear reactions.

C. A **galaxy** is a group of stars and other particles connected by gravity.

D. Our **solar system** consists of planets, asteroids, and other bodies orbiting the Sun.

E. Planet Earth rotates around its axis, causing day and night. The tilt of the Earth on its axis as it revolves around the Sun explains the seasons.

F. The movements of the Earth and the moon are responsible for the phases of the moon that we observe from Earth.

OUR UNIVERSE

The **universe** is everything that exists — all matter, all energy, and all space. The universe extends in every direction as far as we can detect. It may go on forever, or it may in fact be limited in size. Scientists have no way to tell for sure. Most scientists believe that the present universe is expanding; many scientists also believe the universe began in one place at a single point in time almost 15 billion years ago. Today, our universe consists of galaxies of stars, clouds of cosmic gas and dust, and enormous distances of empty space. Scientists measure these large distances by **light years** — the distance that light can travel in one year.

STARS

Stars are enormous balls of superheated gases. Scientists use powerful telescopes to study them. By analyzing the light rays emitted by a star, scientists are able to tell its composition. Most stars are mainly made up of hydrogen and helium. Each star is actually like a giant nuclear reactor, producing energy through **nuclear fusion** — the joining together of atomic nuclei to make new atoms.

FORMATION

Scientists believe that stars first form out of clouds of gases and dust in space known as a **nebula**. The force of gravity pulls the gas and dust together. As the cloud becomes more concentrated, it also begins to spin. Small particles come closer and closer together, until they form a star.

A nebula, 900,000 light years across and 500,000 times the mass of the sun.

The center of a star is extremely hot and dense. Tremendous pressure causes single protons (*hydrogen nuclei*) to fuse together into helium. This process unleashes tremendous amounts of energy. Energy moves from the center of the star outwards. The energy finally radiates from the surface of the star across open space at the speed of light.

LIFE CYCLE OF A STAR

Eventually, a star will turn all its hydrogen into helium. Without the pressure of nuclear fusion pushing the star outward, gravity

The star-studded galaxy M81, a nearby galaxy similar to our own

then causes the star to contract. Next, the internal temperature of the star rises, causing it to expand again and become a **red giant**.

Gradually, the star begins to cool down. When fusion stops and a large star collapses into itself, it experiences a violent explosion. Heavier elements are formed at this time. Eventually, the star cools down again and becomes a **dwarf** or a dense **black hole** — an area of dense matter that attracts other matter by its strong gravitational force.

CHAPTER 13: THE UNIVERSE

> **APPLYING WHAT YOU HAVE LEARNED**
>
> ✦ Describe how stars are formed and how they produce energy.
>
> ✦ Make a timeline showing the life cycle of a typical star.

GALAXIES

The central region of our Milky Way

Stars are grouped together into **galaxies**. A galaxy, like our own **Milky Way**, will often have billions of stars as well as nebulae and black holes. Each galaxy is thousands of light years in width. Galaxies come in many different shapes. Some galaxies, like our own Milky Way, are swirling spirals. Others are shaped like a lens or are simply irregular. Galaxies in turn are grouped into clusters and superclusters. These clusters are giant bands numbering ten thousand to one million galaxies spread across space.

The Sun. The Earth, several other planets, asteroids, comets, and dust orbit the star known as the Sun. Together, these bodies make up what is known as the **solar system** (see the picture on page 154). The Sun is the center of our solar system. It is the largest object in our solar system and contains more than 99% of the solar system's mass. The Sun is so enormous that it could easily hold more than a 1.3 million planets the size of Earth.

THE MOVEMENT OF BODIES IN SPACE

The Role of Gravity. The force of gravity influences the movements of bodies both in space and on Earth. **Gravity** is a force of attraction between any two pieces of matter. The force of gravity increases as the objects move closer together. Gravity also increases along with the masses of the objects. Larger masses will have a stronger gravitational attraction between them than smaller masses that are the same distance apart.

The Movement of the Planets. The force of gravity explains the movement of the Earth and other planets around the Sun, as well as the movement of the moon around the Earth. All the inner planets, for example, circle the Sun in **elliptical** orbits (*shaped like an oval*). At any one time, each planet might move in a straight line and fly off into space, but the force of gravity from the Sun bends each planet's orbit.

Comets. Comets are made of ice, rocks and dust that circle the Sun in long elliptical orbits. As a comet approaches the Sun, some of its ice turns to gas, creating what looks like a giant, glowing tail.

The Moon. Gravity also affects the moon, which revolves around the Earth in a circular orbit. It might travel through space in a straight line, but it is attracted to Earth by the force of gravity. The interaction of these two forces causes the moon to orbit the Earth.

An artist's drawing of the planets as they revolve around the Sun

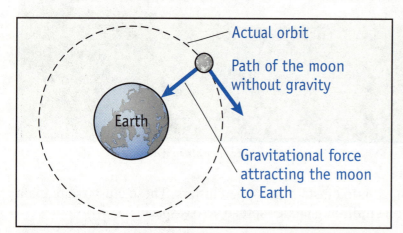

Some scientists believe that the moon came about as a result of a collision between Earth and an object the size of Mars. One theory states that fragments from the collision were hurled into space where gravity brought them together and they fused fo form the moon.

APPLYING WHAT YOU HAVE LEARNED

◆ How would the planets move if the Sun did not exert any gravitational force? Explain your answer.

THE MOVEMENT OF THE EARTH

The planet Earth actually moves in two different ways at the same time: it **rotates** on its axis, and it **revolves** around the Sun.

CHAPTER 13: THE UNIVERSE 155

Earth's Rotation. The Earth **rotates**, or spins, around its **axis** — an imaginary line running through the center of Earth from the North Pole to the South Pole. This rotation takes 24 hours, causing day and night to occur on Earth. Night occurs on those parts of the Earth that are away from the Sun's rays.

Earth's Tilt. The Earth also tilts on its axis. Earth's axis tilts $23\frac{1}{2}°$ away from a line perpendicular to an imaginary line connecting the Earth to the Sun.

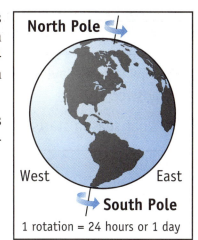

1 rotation = 24 hours or 1 day

EARTH'S REVOLUTION AND THE SEASONS

The Earth revolves around the Sun at the same time that it rotates on its axis. It takes just over 365 days (*one year*) for the Earth to complete one **revolution** around the Sun. Because of its tilt, the Sun's rays hit the Northern Hemisphere longer and more directly in summer than in winter. The Sun appears to rise higher in the sky, temperatures are warmer, and the days are longer. When it is summer in the Northern Hemisphere, it is winter in the Southern Hemisphere. This is because the Southern Hemisphere is tilting away from the Sun and receives less direct solar rays. The area around the Equator is not affected by Earth's tilt. It is always warm because it receives the Sun's direct rays all year. The following chart shows the change of seasons:

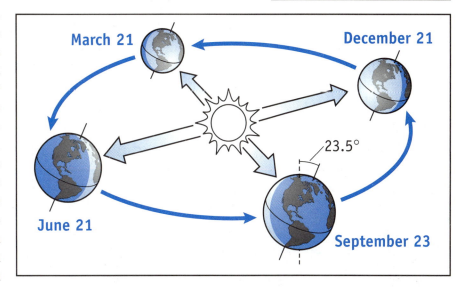

Date	What Happens on that Day
June 21	Longest day of the year in the Northern Hemisphere — the start of summer in that hemisphere
Dec. 21	Shortest day of the year in the Northern Hemisphere — the start of winter in that hemisphere
Mar. 21	Day and night are equal everywhere on Earth — the start of spring in the Northern Hemisphere
Sept. 23	Day and night are equal everywhere on Earth — the start of autumn in the Northern Hemisphere

APPLYING WHAT YOU HAVE LEARNED

◆ How long is daylight in the Southern Hemisphere on March 21? Explain your answer.

◆ What season starts in the Southern Hemisphere on June 21?

LUNAR PHASES

The movements of the Earth and the moon also explain what appears to us as the **phases of the moon**. The amount of the moon we can see each night changes over time in a cycle that repeats itself about once a month. Every 29 days, the moon appears as a thin sliver, grows into a crescent, expands into a full moon, then becomes a crescent again, and eventually narrows until it becomes completely dark in the night sky. These lunar phases are caused by the moon's orbit and the different views from the Earth of the Sun's rays reflected on the moon.

PRINCIPAL PHASES OF THE MOON

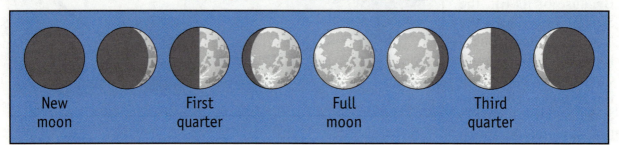

New moon · First quarter · Full moon · Third quarter

The moon orbits the Earth every 29 days. The angle between the Earth and the moon constantly changes. When the Earth and Sun are on the same side of the moon, the entire moon is lit up from Earth, making a **full moon**. When the moon is between the Earth and the Sun, a view from the Earth cannot see any reflection of the Sun's rays from the moon: the **new moon** is totally dark. Because the moon rotates as it orbits Earth, the same side of the moon is always visible from Earth.

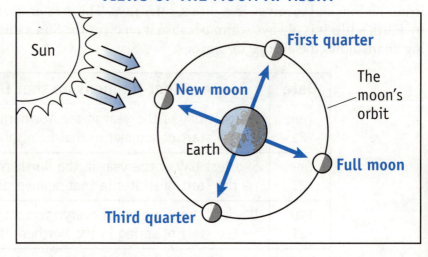

VIEWS OF THE MOON AT NIGHT

APPLYING WHAT YOU HAVE LEARNED

♦ Why does the apparent size of the moon appear to change? Explain your answer.

♦ What explains why the same side of the moon is always visible from Earth?

WHAT YOU SHOULD KNOW

★ You should know that the **universe** consists of all matter, energy, and space.

★ You should know that **stars** are balls of hot gases that produce energy through nuclear reactions.

★ You should know that a group of stars is called a **galaxy**.

★ You should know that gravity is a force of attraction between all forms of matter.

★ You should know that the **rotation** of Earth creates day and night.

★ You should know that the **tilt** of Earth on its axis causes the seasons as Earth revolves around the Sun.

★ You should know that the movements of Earth and the moon lead to **phases** of the moon seen from Earth.

CHAPTER STUDY CARDS

The Movement of Earth

★ **Gravity.** Gravity is a force of attraction between any two objects. Gravity governs the movements of the planets, moons, asteroids and comets in our solar system.

★ **Earth's Movement.** Earth **rotates** on its axis, causing day and night.
- Earth **revolves** around the Sun.
- The **tilt** of the Earth's axis explains the change of seasons as Earth revolves around the Sun.

Stars and the Phases of the Moon

★ **Stars.** Scientists believe stars were formed out of clouds of gases and dust in space known as **nebula**. Stars produce energy through nuclear fusion, converting hydrogen into helium. Stars create all elements besides the lighter gases.

★ **Lunar Phases.** The appearance of the Sun's reflected rays on the moon and the moon's position in its orbit around Earth are responsible for the various phases of the moon.

CHECKING YOUR UNDERSTANDING

1   When the Earth is at its greatest distance from the Sun, its Northern Hemisphere is tilted toward the Sun, as shown in the drawing below.

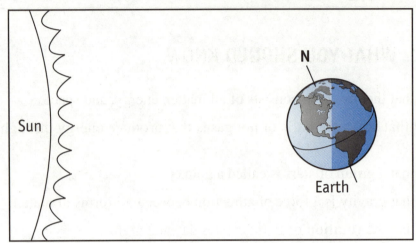

- ◆ Examine the Question
- ◆ Recall What You Know
- ◆ Apply What You Know

What season takes place in the Southern Hemisphere at this time?

A   Spring
B   Autumn
C   Summer
D   Winter

OBJ. 5
7.13 (A)

This question requires you to recall information about Earth's position in space and how this causes the seasons. You should recall that when the Northern Hemisphere tilts towards the Sun, the Southern Hemisphere tilts away. What season does the Southern Hemisphere experience at this time?

Now try answering some questions on your own about astronomy.

2   The time required for the moon to show a complete cycle of phases when viewed from Earth is —

F   24 hours
G   7 days
H   29 days
J   365 days

OBJ. 5
7.13 (B)

3   The daily path of the Sun, as viewed from the Earth, changes with the seasons because —

A   Earth's axis is tilted
B   Earth's distance from the Sun changes
C   the Sun revolves
D   the Sun rotates

OBJ. 5
7.13 (A)

4 The moon's phases can be observed from Earth because the moon —

    F   reflects the Sun's light
    G acts as a nuclear reactor
    H fuses hydrogen nuclei into helium
    J   casts a shadow on the Sun's surface

♦ Examine the Question
♦ Recall What You Know
♦ Apply What You Know

OBJ. 5
7.13 (B)

5 The shaded portion of the map below indicates areas of night, and the unshaded portion indicates areas of daylight.

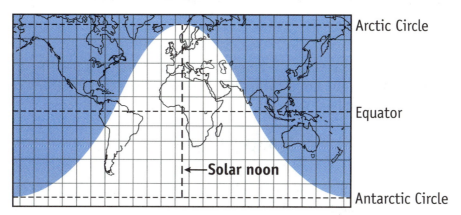

What day of the year is represented by the map?

    A March 21   C December 21
    B June 21   D September 21

OBJ. 5
7.13 (A)

6 On which two dates do all locations on Earth have equal hours of day and night?

    F September 23 & December 21   H March 21 & June 21
    G December 21 & March 21   J March 21 & September 23

OBJ. 5
7.13 (A)

7 Which group is arranged in order from smallest to largest?

    A Earth, moon, and Sun   C Milky Way, Sun, Earth
    B universe, Sun, galaxy   D planet, galaxy, universe

OBJ. 5
8.13 (A)

Use the diagram below to answer the following question.

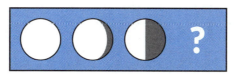

OBJ. 5
7.13 (B)

8 Which of the following is the next phase of the moon?

    F   G   H   J

CHAPTER 14

PLANET EARTH: CYCLES, SYSTEMS AND INTERACTIONS

Our home, the Earth, is the only place we know of in the universe that has life. In this chapter, you will learn about the unique conditions that make our planet habitable. You will learn about the Earth's features and how its crust, oceans, and atmosphere interact.

MAJOR IDEAS

A. Gradual changes, including tectonic plate movement, weathering, and erosion, are responsible for the Earth's land features.

B. Many **cycles** exist in Earth's systems, including the rock cycle, the water cycle, the carbon cycle and the nitrogen cycle. Matter and energy often interact in these cycles.

C. Interactions between the Earth's oceans, atmosphere, and solar energy lead to common weather patterns and other effects.

D. The presence of life has greatly affected the Earth's systems, including the water, carbon and nitrogen cycles. Human activity now affects many of Earth's resources, including the quality of its soil, water and air.

E. Catastrophic and other natural events as well as human activities have contributed to the extinction of many species.

PLATE TECTONICS AND EARTH'S SURFACE FEATURES

At the center of the Earth is an iron core. This core is surrounded by a layer of hot, dense rock that is almost 3,000 km thick. All this is surrounded by the outermost layer of the Earth — a thin skin known as the **crust**. It is on this topmost layer that all life exists on Earth.

PLATE TECTONICS

Earth's crust and part of the hot, dense rock below it are actually divided into several large slabs, known as **tectonic plates**. The map on the right shows where the major plates are located. Each tectonic plate is about 100 km (*60 miles*) thick. These giant plates act like massive solid chunks, floating on top of hot, plastic-like rock.

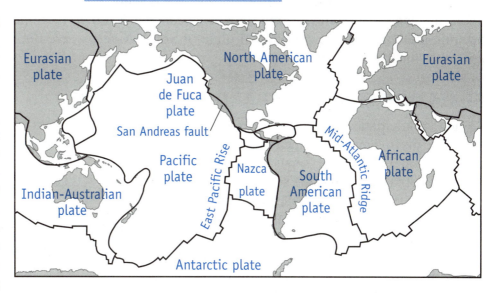

Heat from below the Earth's surface causes the rock inside Earth to move. These movements cause the plates above to shift and slide. Gravity also plays a role in moving these plates. Gravity pulls the heavier parts of the plates downward. Each plate moves very slowly — at a speed of 1 to 16 cm per year. Over hundreds of millions of years, however, a tectonic plate may move thousands of kilometers.

APPLYING WHAT YOU HAVE LEARNED

✦ What role does gravity play in tectonic plate movement?

EFFECTS OF PLATE TECTONIC MOVEMENT

Tectonic plates push and pull against each other like bumper cars in an amusement park. These movements are responsible for some of Earth's major land features.

MOUNTAIN BUILDING

When two land plates, known as **continental plates**, slowly push into one another, they often **fold** upwards, creating mountain chains. The Indian plate, for example, pushes northward against the Eurasian plate. The folding of these two plates has created the Himalayan Mountains — the world's highest mountain chain.

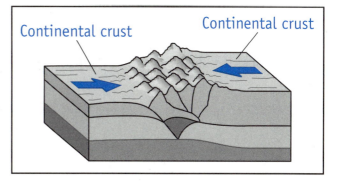

The Earth's crust under the oceans is thinner but denser than the continental crust. When oceanic crust collides into continental crust, the oceanic crust sinks downwards and slides under the continental crust, lifting it up. This can also build up mountains. For example, the oceanic plate lifts up the continental plate of South America, creating the mountain range known as the Andes Mountains.

SEAFLOOR SPREADING AND RIFT VALLEYS

Some tectonic plates move apart. Scientists have discovered that in the middle of the Atlantic Ocean, the separation of plates is actually causing the seafloor to spread. As the plates move apart, magma rises up through the cracks in the ocean floor, creating a ridge of mountains. (*See the Mid-Atlantic Ridge on the previous page*). In other areas, the separation of tectonic plates has created rift valleys — long valleys between parallel ridges of mountains. This creation of new crust would increase the Earth's size, except that it is balanced by the folding and colliding of plates elsewhere.

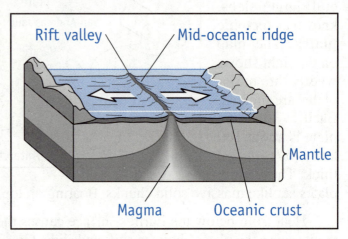

VOLCANOES

A volcano is an opening in the Earth's surface that lets out molten rock and gases. Volcanoes often occur at the edges of tectonic plates. Part of a plate has sunk into the Earth, where it creates molten rock known as **magma**. The magma escapes through any weaknesses in the Earth's crust.

Once the molten rock comes to the surface, it becomes known as **lava**. As lava accumulates on the ground around the volcano, it gives the volcano its typical shape. Many islands and mountains have been formed by volcanoes. For example, the Hawaiian Islands are actually the tops of volcanoes in the Pacific Ocean.

Mount St. Helens in Washington State erupts.

EARTHQUAKES

As tectonic plates move, they often create great stress in the surrounding rock. Sometimes, two plates slide along the same line in opposite directions. Stress continues to build in the rock. Eventually, the rocks release the energy created by this stress in an **earthquake**. The rock vibrates to release this stress. This energy passes through the Earth to the surface in a series of **seismic waves**. Earthquakes most often occur at the edges of tectonic plates. This type of catastrophic event can kill life forms and change rock formations.

> ### APPLYING WHAT YOU HAVE LEARNED
>
> ◆ Explain how the actions of tectonic plates have helped to shape the Earth's land features.

WEATHERING AND EROSION

Tectonic plate movements build mountains through folding. They also create new crust when they separate and new magma comes pouring through. The processes of weathering and erosion reduce mountains and other land features created by volcanoes, earthquakes, and folding.

- ★ **Weathering.** The wearing down of rocks at the Earth's surface by the actions of wind, water, ice and living things is referred to as weathering. Water, for example, expands when it freezes. Water may seep into cracks or pores in rocks and expand these cracks if the temperature drops and the water freezes. Rain and running water will also break down rock into smaller particles. Some chemicals, like acids, will dissolve rocks. Microscopic organisms may also cause rocks to break down and disintegrate.

- ★ **Erosion.** The process by which soil and rock are broken down and moved away is known as erosion. Once rock is broken into smaller particles, wind, running water, glaciers or gravity may cause this sediment to move to a new location. For example, if you have ever visited a sandy beach on a windy day you can understand the abrasive power of wind-driven sand. Rivers likewise carry sediment downstream and deposit this sediment where it meets the ocean. The action of ocean waves can wear down a rocky shoreline or cause the sand in a beach to move into the ocean, causing beach erosion.

Shore erosion at Montara, California, 1998

- ★ **Land Subsidence.** When an area of the Earth's surface weakens and collapses inward, this is referred to as land subsidence. Human mining activities may also lead to subsidence. Often, the depressed area fills with sediment.

EARTH'S CYCLES

Many of the processes of Earth's surface are **cycles** — processes that go through a series of steps in which the last step leads back to the first step, beginning the process all over again. Once a scientist has analyzed the steps of a cycle, he or she can often predict what will happen next in the cycle.

THE ROCK CYCLE

The **rock cycle** provides an example of this process. Rocks are often classified into three groups based on the way they are formed. Cooled magma forms **igneous rock**, such as granite or basalt. Weathering and erosion from water and air breaks down rocks on Earth's surface into pebbles, sand and dust. These fragments pile up and become compressed or cementered together into **sedimentary rock**, like sandstone. Changes from tectonic plate movements may bring sedimentary and igneous rocks below Earth's surface. A great amount of heat and pressure can change these rocks into **metamorphic rock**, such as marble or slate, or even melt the rock completely, so that it forms new igneous rock.

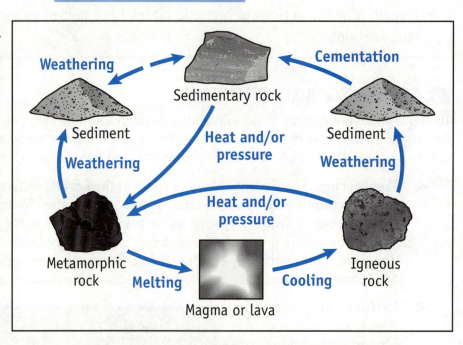

THE WATER CYCLE

One of the most important cycles on Earth is the **water cycle**. The water cycle is the process by which Earth's water moves into and out of the atmosphere. Most of Earth's surface, more than 70 percent, is covered by water. About 97 percent of this water is in the ocean; most of the rest is frozen in the polar ice caps. Less than one percent is found in the Earth's atmosphere, groundwater, lakes and rivers.

The water cycle begins when solar energy, radiating from the Sun, heats the surface of the oceans. This transfer of energy causes some of the surface water to evaporate into the atmosphere.

Plants also create water vapor through **transpiration** and animals through **perspiration**. Water vapor from all these sources rises until it becomes cooler. The water vapor then condenses into droplets small enough to float in the atmosphere as **clouds**. When the droplets grow larger and heavier, they fall back to Earth's surface as **precipitation** — rain, snow or hail. Most precipitation returns to the ocean, but some falls on land, where it is either absorbed by the ground or forms lakes, streams and rivers. Water absorbed by the ground sinks until it hits dense rock, where it collects as **groundwater**. The region in which all groundwater and surface water is collected is known as a **watershed**. Some of the surface water evaporates, but most of the water in the watershed will eventually drain through streams, rivers, and groundwater flows into the ocean.

THE WATER CYCLE

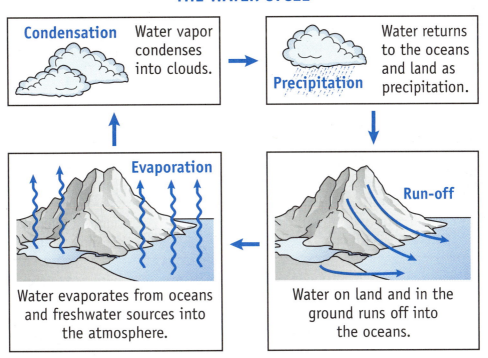

There is a significant interaction between matter and energy in the water cycle. Energy is required for water to evaporate and become water vapor. This energy comes from the Sun. Because energy is absorbed by the water molecules, the process of **evaporation** cools surrounding surface areas and the atmosphere. When water condenses, it releases heat energy back into the air.

APPLYING WHAT YOU HAVE LEARNED

◆ Explain how the rock and water cycles affect the Earth.

THE CARBON AND NITROGEN CYCLES

Living organisms also make up an important part of the Earth and affect its processes. For example, the Earth's primitive **atmosphere** had no oxygen. Over millions of years, early bacteria and primitive plant life absorbed carbon dioxide and created oxygen through photosynthesis. Today, 20% of the Earth's atmosphere is oxygen. Living organisms also affect the recycling of carbon and nitrogen through the Earth's systems.

THE CARBON CYCLE

Carbon is an essential element of life on Earth. It is found in all *organic compounds*, which are the building blocks of life. Carbon is present in the atmosphere in all living things, in layers of limestone on the ocean floor, and in fossil fuels. It is continuously recycled between the atmosphere (*as carbon dioxide or CO_2*), plants, and animals:

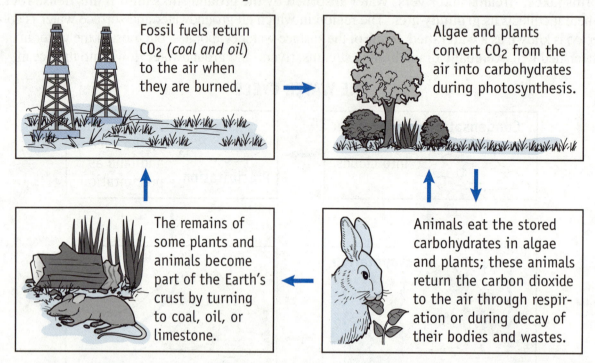

Like the water cycle, the carbon cycle involves interactions between energy and matter. Think of the decay of dead plants and animals (*biomass*). In nature, worms, ants, bacteria and other decomposers break down this biomass into water, minerals, organic compounds and carbon dioxide. In the process, they release some of the energy stored in these carbohydrate molecules. A compost bin is a place where a gardener places decaying biomass to make fertilizer to enrich the soil. The bin speeds up the decaying process. Decomposers release energy as they break down the biomass so that the compost bin actually becomes warm.

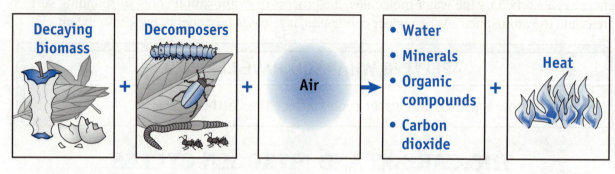

THE NITROGEN CYCLE

Life on Earth also depends on the availability of nitrogen. Like the rock, water, and carbon cycles, the nitrogen cycle reuses the same material in various forms. Although nitrogen is plentiful in the air, this nitrogen is not in a form that animals or plants can use.

Instead, certain bacteria must turn this nitrogen into useful **nitrates**.

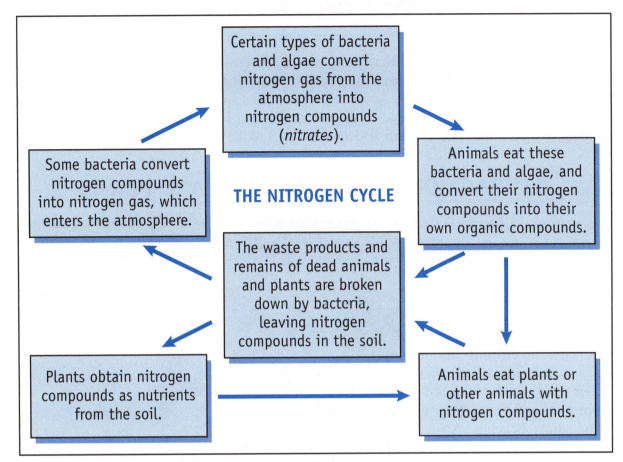

Because different forms of life depend on the water, carbon and nitrogen cycles, significant risks arise when human activities interfere with these cycles. For example, by burning fossil fuels such as coal, humans release more carbon dioxide into the air than plants can convert into carbohydrates. At the same time, humans are cutting down tropical rainforests, reducing the number of plants conducting photosynthesis. When this happens, the carbon cycle is modified and excessive carbon dioxide builds up in the atmosphere.

APPLYING WHAT YOU HAVE LEARNED

✦ Identify three results that might occur from modifying the Earth's water, carbon and nitrogen cycles.

Earth's Cycle	A Possible Result
1. Water Cycle	1.
2. Carbon Cycle	2.
3. Nitrogen Cycle	3.

SYSTEM INTERACTIONS

In Unit 5, you learned about the characteristics of **systems** and how human organ systems **interact**. Just as the human body has different systems, so does Planet Earth. The movement of tectonic plates to shape the Earth's surface features makes up one such system. The gases surrounding the Earth, known as the **atmosphere**, make up a second system. The oceans of the Earth make up a third system. Earth also acts as a part of the larger solar system.

INTERACTION OF OCEANS AND LAND FORMS

These systems often interact. The world's oceans interact with its land forms in a variety of ways. Rivers carry sediment, including salts, into the ocean. Most of the ocean floor is covered by this sediment, which has taken millions of years to accumulate. Ocean currents carry some of this sediment to coastlines, where it forms sandy beaches. At the same time, the actions of tides and waves can erode shorelines, sweeping away sand, chipping away at rock, and dissolving minerals.

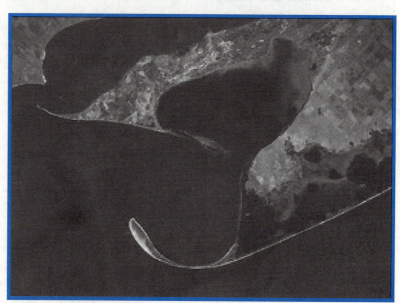

The interaction between the ocean currents and erosion helped create this unique landform.

WEATHER PATTERNS

The **weather** refers to conditions in the atmosphere at the Earth's surface — including temperature, precipitation, humidity, and winds. The energy behind changes in the weather comes from the Sun. **Climate** is the average weather conditions in a place over a long period of time. Weather results from the interaction of several systems — land features, energy from the sun heating the Earth's atmosphere, the spinning of the Earth creating winds, and the conditions of the oceans. For example:

★ Temperatures are generally warmer the closer one gets to the equator. Temperatures become cooler at higher altitudes, such as on mountains or high plateaus.

★ Because air is cooled as it rises above a mountain, the ocean side of a mountain often has heavy precipitation. The air loses moisture and becomes drier by the time it reaches the other side of the mountain.

★ Differences in the **specific heat** of land and the ocean also affect climate. Water requires more energy to change its temperature — so it stays cooler than land in the summer and warmer in winter. This affects air flowing over these areas.

WINDS AND TORNADOES

The spinning of Earth and the uneven heating of the atmosphere by the Sun create wind patterns. Cold sinking air creates areas of high pressure, while hot rising air creates low pressure. Winds blow from high to low pressure areas. Tornadoes are high-speed winds that whirl in a funnel. A tornado occurs when dry, cool air meets warm, humid air. The warm air rises quickly, sucking in both air and objects.

TROPICAL HURRICANES

Hurricanes occur in tropical regions in late summer and early fall when the Sun has heated the ocean water to very warm temperatures. The warm ocean water evaporates so quickly that it creates an area of low pressure. Air around the rising air column begins to spiral at high speeds. The hot air rises until it cools and condenses, releasing energy and causing heavy rains, winds and lightning.

A tropical hurricane approaches the coast of Mexico

APPLYING WHAT YOU HAVE LEARNED

◆ Describe two interactions among solar, weather and ocean systems.

EARTH'S CHANGING PROCESSES

Sometimes, the delicate balance in Earth's cycles and systems can be upset. For example, a catastrophic natural event like the crashing of a meteor onto the Earth's surface, an earthquake, a volcanic eruption, a fire, or a change in climate may completely transform a region or ecosystem.

Some events may even change the Earth's surface. A volcanic eruption can create a mountain. Other catastrophic events will mainly affect life forms. A bolt of lighting might start a fire that burns down an entire forest. All the animals that depended on the forest for food and shelter would probably die. Then a succession of ecological changes would follow.

When catastrophic events harm a species in all its locations, the species may even go extinct. Such events can be natural or the result of human activity. For example, many scientists believe a large meteor crashed into the Earth about 65 million years ago. The crash created immense clouds of dust that filled the atmosphere and blocked out much of the visible light. Plants died and dinosaurs, which required large amount of plants for food, were not able to survive. All the species of dinosaurs became extinct in a short period of time.

ENVIRONMENTAL ISSUES

Today, human activities threaten many of Earth's natural processes and the survival of many species. The following are some of the problems that have resulted from human activity:

POLLUTION

The rise of industry in the past two centuries has led to a great decline in air and water quality. Exhaust from cars and factories as well as liquid and solid wastes from manufacturing and urban centers cloud the air and pollute water supplies. Oil spills cover parts of the ocean and shoreline. Since almost all living organisms depend upon clean air and water, pollution poses a severe threat to the survival of all forms of life on Earth. In addition to this general threat, pollution poses several special dangers:

★ **Global Warming.** The burning of fossil fuels like coal and oil (*gasoline*) has significantly increased amounts of carbon dioxide in the atmosphere. Carbon dioxide and water act together to wrap the planet in a "blanket," holding in heat. With increased amounts of carbon dioxide, less heat is able to escape, leading to the "**greenhouse effect**." Winters are milder and summers are hotter. If temperatures continue to rise, part of the polar ice caps could melt and sea levels would rise.

★ **The Ozone Layer.** Free oxygen combines with oxygen molecules to create **ozone** in the Earth's upper atmosphere. This ozone absorbs much of the sun's ultraviolet radiation. Without an ozone layer, ultraviolet radiation would cause mutations in most living cells. The use of chlorofluorocarbons as coolants in refrigerators and air conditioners threatens the ozone layer. Each CFC molecule can break down thousands of ozone molecules. As a result, an ozone "hole" has appeared in the Earth's atmosphere, leading to an increased incidence of skin cancer. Countries have agreed to ban CFCs, although some still use them.

★ **Pesticides.** Poisonous chemicals are used to kill insects that threaten crops, but these pesticides then become part of the water and soil, endangering other organisms as well, such as birds. These pesticides may even be absorbed by the crops we grow for food.

★ **Acid Rain.** When coal and oil are burned, they dump pollutants into the atmosphere. Many pollutants released by industry and automobile exhausts turn into acids. These acids get washed out of the air when it rains. When these pollutants return in rain water, they are highly toxic, killing fish, destroying forests, eroding soil and further endangering the environment.

LOSS OF NATURAL RESOURCES

Some resources, like underground water, can renew themselves after a period of time. These are known as **renewable resources**. Other resources, like oil and coal, are **non-renewable**, and can only be used once. A third group of resources, like nitrogen gas, are **inexhaustible**. Many human activities, like the burning of fossil fuels, are rapidly using up Earth's non-renewable resources. Other activities are using renewable resources, like trees, at a faster rate than they can renew themselves.

DESTRUCTION OF NATURAL HABITATS

Destruction in the Amazon rain forest

One of the greatest threats to the environment is the destruction of many natural habitats. As the human population expands, more and more forests, wetlands, and grasslands are destroyed to build farms, factories, and cities. The destruction of tropical rain forests is one of the most dramatic examples of the loss of natural habitats. Tropical rain forests have the greatest **biodiversity** (*diversity of species*) and the greatest concentration of plant life. The destruction of areas like the Amazon rain forest reduces the amount of oxygen in the atmosphere and leads to the **extinction** of many species. More plant and animal species now become extinct each year than at any other time since the extinction of dinosaurs. This is especially important, since genetic material in some of the species going extinct may contain cures for many diseases.

When a species comes close to extinction, it becomes known as an **endangered species**. In the United States, government agencies list endangered species. Once a plant or animal is identified as endangered, it becomes unlawful to kill it, and even its natural habitat is protected.

APPLYING WHAT YOU HAVE LEARNED

◆ What conclusions can you make about the effects of human activity on Earth's renewable, non-renewable, and inexhaustible resources?

◆ Describe how human activities have modified Earth's soil, water, and air quality.

WHAT YOU SHOULD KNOW

★ You should know that gradual changes caused by tectonic plate movement, weathering, erosion, and land subsidence are responsible for many of the Earth's land features. For example, the collision of tectonic plates can cause upward folding, creating mountain chains; volcanoes also build mountains.

★ You should know that **cycles** exist in Earth's systems — the rock cycle, water cycle, carbon cycle and nitrogen cycle. Matter and energy often interact in these cycles. For example, decomposers break down the remains of plants and animals as part of the carbon cycle, releasing energy.

★ You should know that different Earth **systems**, such as the solar, ocean, and weather systems, frequently interact. For example, solar energy can warm ocean water very rapidly in tropical regions. Water vapor may then rise so quickly that a tropical hurricane develops.

★ You should know that the delicate balance of Earth's systems can easily be disturbed. Catastrophic events, like a forest fire or meteor crash, may even lead some species to become extinct. Human activities, such as pollution and damage to the ozone layer, now threaten many of Earth's systems. Humans also threaten to exhaust many **non-renewable resources**.

CHAPTER CARDS

Earth's Land Features

★ **Tectonic Plates.** These are pieces of the Earth's crust and the rock below, about 100 km thick, that slowly move on the Earth's surface. Their movements can create mountains, seafloor spreading, earthquakes and volcanoes.

★ **Weathering.** Wearing down of rock by wind, water, ice and living organisms.

★ **Erosion.** When rock or soil is broken down into pebbles, sand or dust and transported away.

★ **Land Subsidence.** When part of the Earth's surface weakens and sinks.

Earth's Cycles

Cycles can be analyzed and predicted.

★ **Rock Cycle.** Rocks move from igneous to sedimentary to metamorphic and back again.

★ **Water Cycle.** Water evaporates from the ocean and other surfaces. Water then condenses into clouds and later falls back to the ground as precipitation. Ground and surface water collect in a watershed and drain off into the ocean.

★ **Carbon and Nitrogen Cycles.** Involve living things: for example, plants create organic compounds; animals eat plants; carbon is released from their remains, wastes and respiration.

CHAPTER 14: PLANET EARTH: CYCLES, SYSTEMS AND INTERACTIONS

The Interaction of Earth's Systems

★ **Earth's Systems.** Earth's systems often interact. For example, solar energy and water from the oceans interact with the atmosphere to create weather patterns.

★ **Influence of Disastrous Events.** Catastrophic events, like a meteor crash, can lead to the extinction of an entire species. **An endangered species** is protected by government agencies because it is close to extinction.

Impact of Humans on Earth's Systems

★ **Environmental Problems:**
  • **Global Warming.** Burning of fossil fuels has increased carbon dioxide in the air.
  • **Ozone Layer** absorbs much of the sun's ultraviolet radiation, but is being destroyed by CFCs.
  • **Pesticides.** Pesticides can poison water, soil, and the food we eat.
  • **Acid Rain.** Air pollutants turn into acids that are highly toxic.

★ **Loss of Non-Renewable Resources**
★ **Destruction of Natural Habitats**

CHECKING YOUR UNDERSTANDING

1. **Frequent earthquakes and volcanic eruptions occur in an area of the Pacific Ocean called the "Ring of Fire." Which of the following leads to these events?**

   A Earth's rotation on its axis
   B The formation of canyons
   C Movement of tectonic plates
   D Rise in ocean surface temperatures

   ◆ Examine the Question
   ◆ Recall What You Know
   ◆ Apply What You Know

   OBJ. 5
   8.14 (A)

HINT: This question requires you to recall information about natural events that affect Earth's land features. You should recall that tectonic plate activity often leads to the creation of major surface features of the Earth — such as mountains, earthquakes and volcanoes.

Now try answering some questions on your own about
Earth's cycles, systems and interactions.

2. **Which is the best example of weathering?**

   F The cracking of rocks by the freezing and melting of water
   G The transportation of sediment in a stream
   H The condensation of water vapor into droplets in clouds
   J The formation of a sandbar along the side of a stream

   OBJ. 5
   7.14 (C)

3 Which gas in the atmosphere would decrease if a large area of tropical rainforest were planted?

   A Nitrogen
   B Oxygen
   C Carbon dioxide
   D Hydrogen

   ♦ Examine the Question
   ♦ Recall What You Know
   ♦ Apply What You Know

   OBJ. 5
   8.12 (C)

Igneous

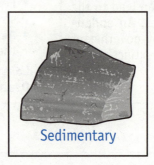
Sedimentary

Metamorphic

4 Rocks are classified as igneous, sedimentary, or metamorphic based primarily on their —

   F texture
   G crystal or grain size
   H method of formation
   J mineral composition

   OBJ. 5
   8.12 (A)

5 As the amount of precipitation on land increases, the level of the groundwater will probably —

   A fall
   B remain the same
   C rise
   D fall and then rise

   OBJ. 5
   6.14 (B)

Read the following paragraph to answer question 6

The Earth's climate is in a delicate state of balance. Even a small change in the factors affecting climate may lead to a catastrophic cooling or warming of the Earth's atmosphere over a period of time. During the last 100 years, measurements have shown a gradual increase in the carbon dioxide in the atmosphere. This change has been linked to an increase in the atmosphere's average temperatures.

6 The scientist who wrote this paragraph most likely believes that —

   F the amount of carbon dioxide in Earth's atmosphere is decreasing
   G recent changes in average global temperatures will have no impact on ocean water levels
   H any change in existing conditions may severely affect the Earth's climate
   J the warming of Earth's atmosphere is largely beneficial

   OBJ. 5
   7.14 (A)

7   The most frequent cause of major earthquakes is the —

   A   sliding of tectonic plates
   B   occurrence of landslides from precipitation
   C   gravitational force between Earth and the moon
   D   change in underwater currents

   OBJ. 5
   8.14 (A)

8   Mountain building rock structures like the photograph to the right are most often caused by —

   F   tectonic plate collisions
   G   the movement of a glacier
   H   heavy rainfall in tropical regions
   J   deposits of sediment

   OBJ. 5
   7.14 (A)

9   A scientist wishes to investigate the interaction of matter and energy during the decay of biomass in a compost bin. Which would be the best experimental design to investigate their interaction?

   A   Observe the color of the compost at regular time intervals.
   B   Examine the odor of the compost bin at regular time intervals.
   C   Measure the temperature of the compost bin at regular time intervals.
   D   Measure the humidity of the compost bin at regular time intervals.

   OBJ. 5
   6.8 (B)

10   Which of the following is an example of a renewable resource?

   F   Uranium
   G   Trees
   H   Oil
   J   Coal

   ♦ Examine the Question
   ♦ Recall What You Know
   ♦ Apply What You Know

   OBJ. 5
   7.14 (C)

11   A scientist wishes to create a model of the rock cycle. She uses modeling clay to represent igneous rock. She then breaks the modeling clay into tiny pieces and presses these pieces together to represent sedimentary rock. How could she best represent metamorphic rock?

   A   Break the clay into smaller pieces again.
   B   Mix the pieces of clay with water.
   C   Heat the pieces of clay in an oven while pressing them.
   D   Heat and mold the clay together into a single uniform piece.

   OBJ. 5
   8.12 (A)

Use the diagram below to answer question 12

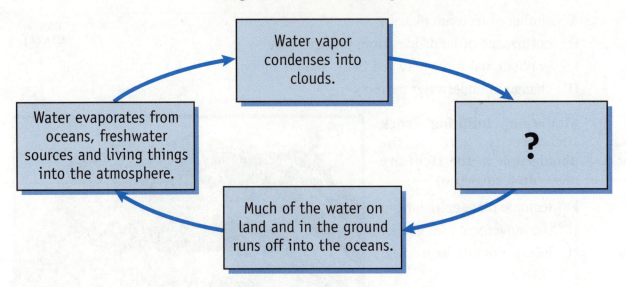

12 Which step is missing from this diagram of the water cycle?

    F Changes in air pressure create tornadoes.
    G The gravitational force of the moon creates ocean tides.
    H Water returns to the oceans and land as precipitation.
    J Water is trapped as ice in the polar ice caps.

OBJ. 5
8.12 (C)

13 The weather map above indicates a coming change in the weather of the Midwest region. The energy that causes these weather changes comes from —

    A tectonic plate movement
    B the conversion of electrical energy
    C geothermal forces below Earth's surface
    D the radiation of the Sun

OBJ. 5
8.10 (B)

14 Which human activity would be more likely to have a negative impact on the environment than the other three?

    F developing research aimed toward the preservation of endangered species
    G using insecticides to kill insects that compete with humans for food
    H using reforestation and cover cropping to control soil erosion
    J investigating the use of biological controls for pests

OBJ. 5
8.14 (C)

CHECKLIST OF OBJECTIVES IN THIS UNIT

Place a check mark (✔) next to those objectives you understand. If you have trouble recalling an objective, review the chapter in the brackets. **You should be able to:**

- ☐ describe characteristics of the universe such as stars and galaxies. **[Chapter 13]**
- ☐ explain how the tilt of the Earth on its axis as it rotates and revolves around the Sun causes changes in seasons and the length of a day. **[Chapter 13]**
- ☐ relate the Earth's movement and the moon's orbit to the phases of the moon seen from the Earth. **[Chapter 13]**
- ☐ predict land features resulting from gradual changes such as mountain building, beach erosion, land subsidence (including tectonic plate movement). **[Chapter 14]**
- ☐ analyze the effects of erosion and weathering. **[Chapter 14]**
- ☐ analyze and predict the sequence of events in the lunar and rock cycles. **[Chapter 14]**
- ☐ identify relationships between groundwater and surface water in a watershed. **[Chapter 14]**
- ☐ explain and illustrate the interactions between matter and energy in the water cycle and in the decay of biomass, such as in a compost bin. **[Chapter 14]**
- ☐ predict the results of modifying the Earth's nitrogen, water, and carbon cycles. **[Chapter 14]**
- ☐ describe interactions among solar, weather, and ocean systems. **[Chapter 14]**
- ☐ describe and predict the impact of different catastrophic events on the Earth. **[Chapter 14]**
- ☐ analyze how natural or human events may have contributed to the extinction of some species. **[Chapter 14]**
- ☐ make inferences and draw conclusions about effects of human activity on Earth's renewable, non-renewable, and inexhaustible resources. **[Chapter 14]**
- ☐ describe how human activities have modified soil, water, and air quality. **[Chapter 14]**

UNIT 7: A PRACTICE MIDDLE SCHOOL TAKS IN SCIENCE

This chapter consists of a complete practice **Middle School TAKS in Science**. Before you begin, let's review a few directions for the test:

- ★ **Answer All Questions.** The TAKS in Science consists of 50 multiple-choice questions. Do not leave any questions unanswered, since there is no penalty for guessing. Blank answers are counted as wrong.

- ★ **Use the "E-R-A" Approach.** Remember to carefully examine the question to understand what it is asking. Next, recall what you have learned about that particular topic in science. Finally, apply your knowledge to answer the question.

- ★ **Use the Process of Elimination.** When answering a multiple-choice question, it should be clear that certain choices are wrong. They will be irrelevant, lack a connection to the question, or be inaccurate. After you eliminate incorrect choices, select the best response that remains.

- ★ **Revisit Difficult Questions.** It is possible that you will come across several difficult questions. If you run into a difficult question, do not be discouraged. Circle the question, or put a mark (✔) next to it. Answer it as best you can and move on to the next question. At the end of the test, go back and reread any questions you marked. Sometimes the answer to a difficult question might become clearer to you with a second reading.

- ★ **Read Items Carefully.** Read the directions carefully. Be sure to examine all parts of the question. If you have time left at the end, check your work and correct any errors.

- ★ **When You Finish.** When you finish the test, make sure you have answered all the questions and that your answers correspond to the correct question number. Do not disturb other students!

As with every question in this book, this practice **TAKS in Science** indicates the objective and grade level number tested by each question. This information is provided to help you and your teacher identify any areas you still need to study.

Good luck on this practice test!

CHAPTER 15

A PRACTICE TAKS IN SCIENCE

*For this practice test, use either the **Periodic Table of the Elements** your teacher will provide or the one in this book on page 67. The **Formula Chart** is below. On the actual Middle School TAKS in Science, copy of the Periodic Table of the Elements and a Formula Chart will be provided.*

FORMULA CHART
for Middle School—Grade 8 Science Assessment

Work = force × distance	$W = Fd$
Speed = $\dfrac{\text{distance}}{\text{time}}$	$s = \dfrac{d}{t}$
Force = mass × acceleration	$F = ma$
Weight = mass × acceleration due to gravity	$Weight = mg$
Density = $\dfrac{\text{mass}}{\text{volume}}$	$D = \dfrac{m}{v}$

Constants/Conversions
g = acceleration due to gravity = $9.8 \, \dfrac{m}{s^2}$
speed of light = $3 \times 10^8 \, \dfrac{m}{s}$
speed of sound = $343 \, \dfrac{m}{s}$ at sea level and 20°C
$1 \, cm^3 = 1 \, mL$

1. The diagram to the right shows the cell of an unicellular organism. Which part of the cell contains the genetic information needed for the cell to carry on the functions of life?

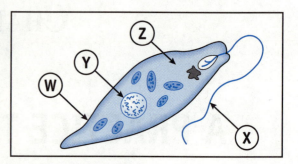

F W
G X
H Y
J Z

OBJ. 2
8.11 (C)

2. Which of the following is an example of a chemical reaction?

A A magnet separates iron filings from sand.
B A kettle of water boils.
C A candle burns on a birthday cake.
D A spoonful of sugar dissolves in water.

OBJ. 3
8.9 (A)

3. Which of the following best represents the structure of a helium (He) atom?

OBJ. 3
8.8 (A)

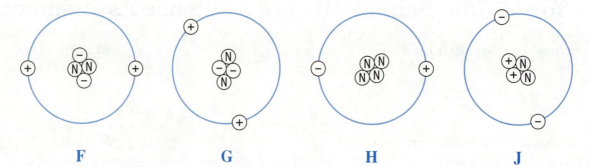

F   G   H   J

4. The diagram to the right represents part of our solar system. The elongated orbit shows the path of an object of rock and ice that forms a glowing tail when it approaches the sun. This object is known as —

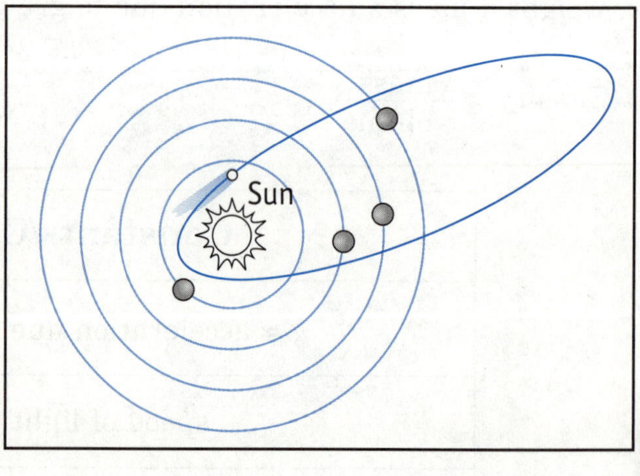

A a meteor
B a planet
C an asteroid
D a comet

OBJ. 5
8.13 (A)

5. Which of the following best explains Earth's four seasons?

F The gravitational pull of the moon
G The tilt of the Earth on its axis as it revolves around the Sun
H The distance between Earth and the sun
J The effect of tectonic plate movement on Earth's oceans

OBJ. 5
7.13 (A)

Use the information below and your knowledge of science to answer questions 6 and 7.

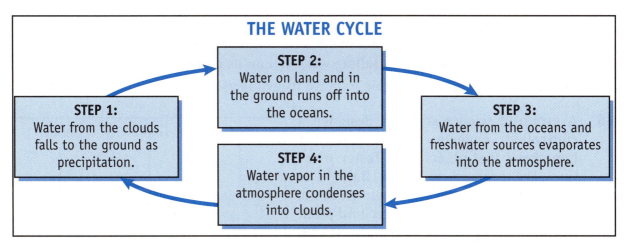

6. Which step of the water cycle, illustrated in the diagram above, makes surrounding areas cooler because it absorbs energy?

   A Step 1
   B Step 2
   C Step 3
   D Step 4

   OBJ. 5
   6.8 (B)

7. What would be the most likely effect on the water cycle if Earth's average temperatures rose because of "global warming"?

   F Greater evaporation and greater precipitation
   G Greater evaporation and less precipitation
   H Less evaporation and greater precipitation
   J Less evaporation and less precipitation

   OBJ. 5
   8.12 (C)

8. If a force of 10 newtons is applied to the right side of the lever shown in the diagram below, how much work is performed in lifting the load shown on the left side of the level? Record and bubble in your answer on the grid below.

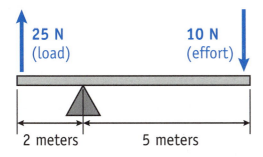

OBJ. 4
7.6(A)

9 Which of the following examples shows a change from chemical to light energy for human use?

A A car battery causes the car's headlights to shine.
B An electric lamp is plugged into a wall socket and turned on.
C A hydroelectric plant uses falling water to turn its turbines.
D A car engine ignites gasoline to move its cylinders.

OBJ. 4
6.9 (A)

10 A group of science students designed a controlled experiment to test the hypothesis that plants grow faster when they are exposed to green light. The design of this experiment is shown to the right. Equal amounts of water and plant fertilizer were given to each plant. Which of the following changes would most improve the experimental design?

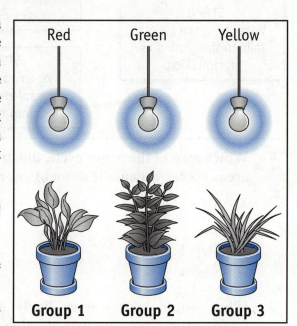

F Expose all three plants to just green light.
G Add a variety of different plants.
H Use the same type of plant in all three groups.
J Use a different size of pot for each plant.

OBJ. 4
6.9 (A)

11 The following diagram shows the relationship of different levels of organization in living organisms:

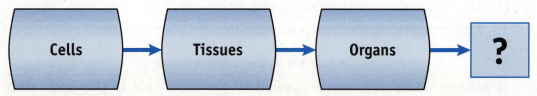

Which of the following would best complete the empty box in the diagram?

A Habitats
B Ecosystems
C Organisms
D Organ Systems

OBJ. 2
6.10 (C)

12 According to its location on the Periodic Table of the Elements, silicon (Si) is classified as a —

F Metal
G Metalloid
H Nonmetal
J Noble gas

OBJ. 3
8.9 (A)

13 In pea plants, the gene for wrinkled seeds (W) is dominant over smooth seeds (w). If a pea plant with one gene for wrinkled seeds (W) and one gene for smooth seeds (w) was crossed with a pea plant with smooth seeds (ww), what percent of their offspring would have wrinkled seeds? Record and bubble in your answer on the grid below.

OBJ. 2
8.11 (C)

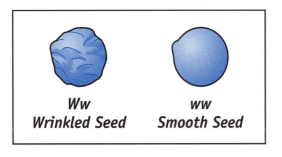

14 Four students were given different tasks to perform as part of a classroom experiment. Which student shown below is not following correct safety precautions?

OBJ. 1
8.1(A)

A

B

C

D

15 The line graph below shows the growth of a certain type of bacteria in a Petri dish for five hours. If trend indicated in the graph continues, what will be the approximate number of bacteria in the Petri dish after six hours?

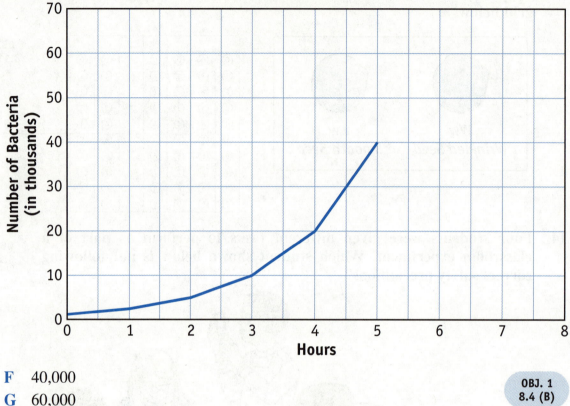

- **F** 40,000
- **G** 60,000
- **H** 80,000
- **J** 100,000

OBJ. 1
8.4 (B)

16 Which of the following is an example of a feedback mechanism?

- **A** When blood sugar (*glucose*) is high, the pancreas produces more insulin to lower the amount of blood sugar.
- **B** The circulatory and respiratory systems interact to carry oxygen to cells throughout the body.
- **C** A plant converts natural sunlight into chemical energy.
- **D** Many traits in living organisms are inherited.

OBJ. 2
8.6 (B)

17 Which evidence best supports the theory that Earth's crust consists of giant shifting plates?

- **F** The Himalaya Mountains are the highest in the world.
- **G** Some volcanoes in the state of Hawaii are still very active.
- **H** Weathering and erosion gradually reduce mountains and other land features.
- **J** Scientists have detected seafloor spreading in the middle of the Atlantic Ocean.

OBJ. 1
8.3 (A)

Use the information below and your knowledge of science to answer questions 18 through 21.

The diagram below illustrates a food web in a pond ecosystem. The ecosystem includes the pond and pond grass growing inside and around the pond.

18   What would most likely occur in this pond ecosystem if a new type of frog were introduced that ate mosquito larvae?

   A   The amount of algae and pond grass would decrease.
   B   The number of yellow perch would increase.
   C   The amount of algae and pond grass would increase.
   D   The number of valve snails would increase.

   OBJ. 2
   8.11 (A)

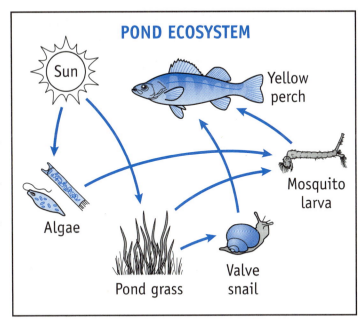

19   From the information in the diagram, yellow perch are best described as —

   F   decomposers
   G   producers
   H   herbivores
   J   carnivores

   OBJ. 2
   7.12 (B)

20   In this pond ecosystem, the pond grass obtain valuable nitrates from the soils. How are these nitrates added to the soil?

   A   They are a byproduct of photosynthesis.
   B   The valve snail produces nitrates during respiration.
   C   Decomposers put nitrates into the soil.
   D   Energy from sunlight puts nitrates into the soil.

   OBJ. 2
   7.12 (B)

21   What process allows the algae and pond grass to convert radiant energy from the sunlight into chemical energy?

   F   Photosynthesis
   G   Respiration
   H   Digestion
   J   Turgor pressure

   OBJ. 2
   8.10 (B)

★★★★★★★★★★★★★★★★

22 Which gas in the atmosphere would increase if a large number of new trees were planted?

A nitrogen
B carbon dioxide
C hydrogen
D oxygen

OBJ. 5
7.14 (C)

23 What is the average length of the three mealworms in centimeters? Record and bubble in your answer on the grid below.

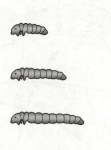

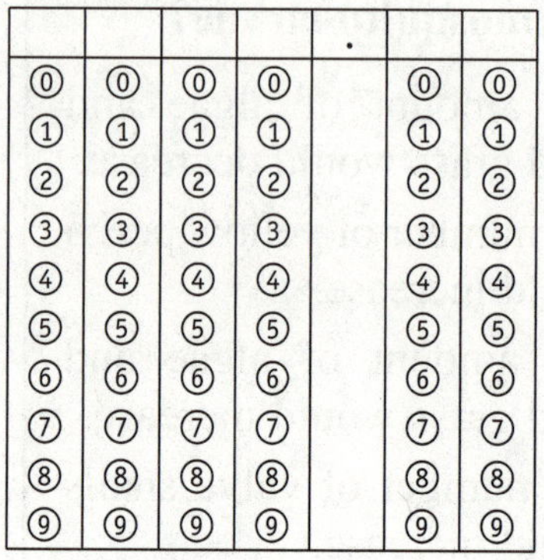

OBJ. 1
8.2 (B)

24 The diagram below represents a hydrogen (H) atom. What step could be taken to improve this model?

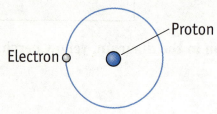

F Increase the size of the electron.
G Increase the distance between the electron and the proton.
H Add additional electrons.
J Add one neutron.

OBJ. 1
8.3 (C)

25 What is the volume of solution in this graduated cylinder in milliliters? Record and bubble in your answer on the grid below.

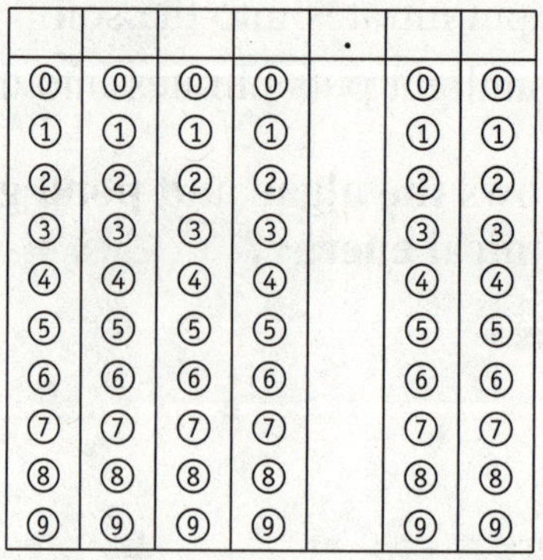

OBJ. 1
8.4 (A)

26 Which of the following is an example of a positive impact of human activity on the environment?

    A The destruction of many natural habitats.
    B The pollution of freshwater sources.
    C The endangerment of some species.
    D The creation of artificial reefs to preserve forms of marine life.

27 A railroad train is moving along its track. At what point on the track does the train have the greatest potential energy?

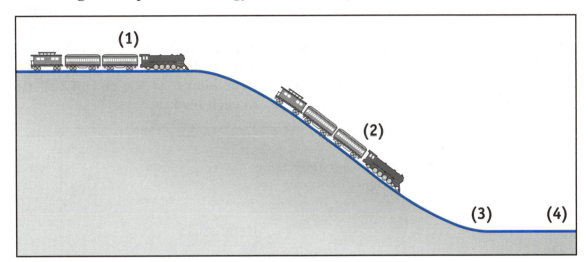

    F Point (1)
    G Point (2)
    H Point (3)
    J Point (4)

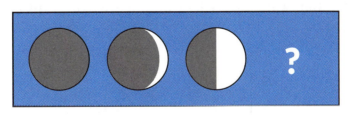

28 Which of the following will be the next phase of the moon?

    A   B   C   D

29 Ozone exists in Earth's upper atmosphere, where it forms a protective layer reducing the impact of the Sun's powerful ultraviolet radiation on Earth's surface. The use of certain refrigerants has created a "hole" in the ozone layer. What is most likely to happen if human activity continues to damage the ozone layer?

F Ultraviolet radiation will cause air conditioners to overheat.
G There will be a shortage of Earth's non-renewable resources.
H Ultraviolet radiation will damage living organisms.
J Ultraviolet radiation may interfere with satellite communications.

OBJ. 5
7.14 (C)

30 José conducts an experiment to see how the heat from a 60-watt light bulb affects the evaporation rate of water. He places 500 mL of water into three identical beakers. Then he places each beaker at a different distance from the light bulb. José measures the distance of each beaker from the light bulb in centimeters. He then puts the light bulb on for exactly three hours. At the end of this time, he measures the volume of water remaining in each beaker.

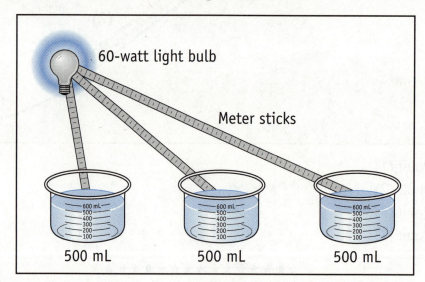

Which of the following would be the best way of graphing the variables in this experiment?

OBJ. 1
8.2 (E)

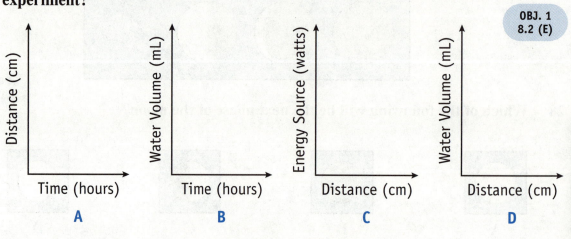

31 The length of a day on Earth is determined by the time required for one complete —

    F rotation of Earth on its axis
    G revolution of Earth around the Sun
    H orbit of the moon around Earth
    J rotation of the Sun

OBJ. 5
7.13 (A)

32 Which of the following illustrations correctly depicts the movement of Earth and its moon?

OBJ. 5
7.13 (B)

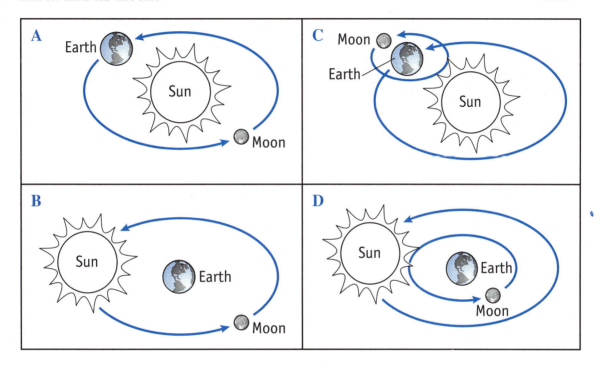

33 A group of students measured the mass and the volume of four liquids. They recorded their results in the table below.

Liquid	Mass (grams)	Volume (cm³)
Corn syrup	10.8	10.0
Water	20.0	20.0
Salad oil	23.0	25.0
Vinegar	30.3	30.0

Which of these liquids had the greatest density?

    F Corn syrup   **H** Salad oil
    G Water   **J** Vinegar

OBJ. 1
8.2 (C)

Use the information below and your knowledge of science to answer questions 34 – 37.

Students place 10g of baking soda (sodium bicarbonate) in a beaker containing 100 mL of water. They stir the mixture until the baking soda is dissolved. Students place 10g of lemon juice into a second beaker containing 100 mL of water, and stir until the juice and water are thoroughly mixed. The students measure and record the temperature of each solution. Then they combine the two solutions and gently stir them together. The students measure and record the temperature of the mixed solution five minutes after the liquids are combined, and again sixty minutes later. The data from the experiment is recorded below.

Solution	Temperature
Baking soda	23°C
Lemon juice	23°C
Combined mixture after 5 minutes	19°C
Combined mixture after 60 minutes	23°C

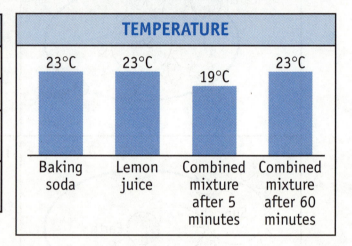

34 Which safety measure should students take in this experiment?

   A   Wear gloves when eating snacks during laboratory time.
   B   Turn hot plates off when experiment is over.
   C   Wear splash-proof safety goggles during the experiment.
   D   Smell each chemical before mixing it with water.

OBJ. 1
8.1(A)

35 What hypothesis is being tested during this experiment?

   F   Combining the solutions will affect their temperatures.
   G   Combining the solutions will affect their color.
   H   Combining the solutions will affect how they smell.
   J   Combining the solutions will affect how they taste.

OBJ. 1
8.2(A)

36 Which laboratory equipment is required for conducting this experiment?

   A   Beakers, graduated cylinders, water test kits
   B   Beakers, thermometer, Petri dishes, test tubes
   C   Beakers, thermometer, graduated cylinders, balance
   D   Beakers, meter stick, hot plate, computer probe

OBJ. 1
8.4(A)

37 **Based on the data collected by the students, what most likely happened when the solutions were combined during this experiment?**

F   Energy was released by a chemical reaction.
G   No change occurred in either the baking soda or lemon juice.
H   Combustion occurred when the baking soda and lemon juice combined with oxygen in the air.
J   The lemon juice and baking soda absorbed energy while combining.

OBJ. 3
8.10(A)

Use the following food chain to answer questions 39 and 40.

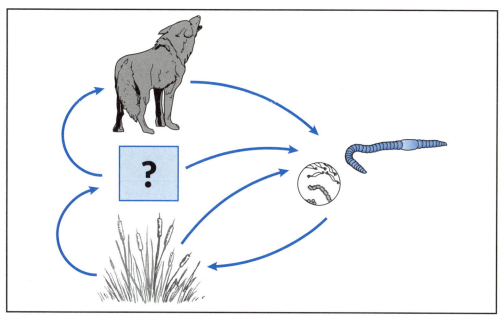

38 **Which of the following best completes the diagram?**

OBJ. 2
7.12 (B)

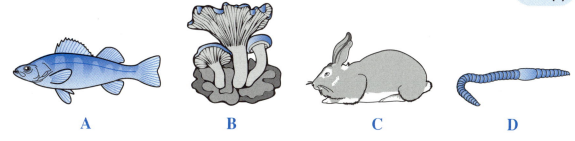

A   B   C   D

39 **The greatest limitation of this model is that —**

F   the nitrogen and carbon cycles are not shown
G   other types of organisms also live in this ecosystem
H   these organisms exist in other ecosystems
J   some of these organisms are endangered species

OBJ. 1
8.3(C)

40 Which two systems interact to send nutrients to cells throughout the body?

A circulatory and nervous systems
B respiratory and digestive systems
C circulatory and digestive systems
D endocrine and excretory systems

OBJ. 2
8.6(A)

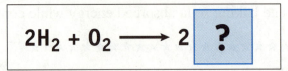

41 The above represents a chemical reaction. What chemical formula fits in the empty box?

OBJ. 3
8.9 (C)

H_2O	CO_2	H_2O_2	NaCl
F	G	H	J

★★★★★★★★★★★★★★★★★★★★★★

42 A ripple tank is a shallow container of water used to demonstrate the properties of a wave. Esmeralda tossed a pebble into the center of the tank. The illustration to the right shows how many waves were generated in 2 seconds. What is the frequency of the waves in the tank?

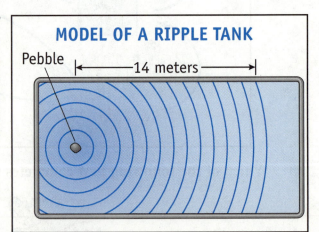

MODEL OF A RIPPLE TANK
Pebble — 14 meters

A 1/7 meters per second
B 7 waves per second
C 14 waves per second
D 1/14 meters per second

OBJ. 4
8.7(B)

43 A man bends his arm at the elbow to lift a heavy book from his desk. In this example, the man's arm is acting as —

F a pulley
G an inclined plane
H a set of gears
J a lever

OBJ. 4
8.7(C)

44 Soil organisms, such as fungi, worms and bacteria, are all parts of a woodland ecosystem. What best describes the role of these organisms in the woodland ecosystem?

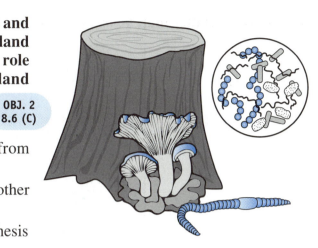

OBJ. 2
8.6 (C)

A To provide oxygen to animals
B To obtain dissolved oxygen from moisture
C To break down the remains of other living things
D To store chlorophyll for photosynthesis

45 Which best explains why some species have become extinct?

F They adapted to their environment through genetic change and natural selection.
G Environmental changes occurred faster than these species were able to adapt.
H Their genes did not allow for change.
J They did not have any recessive traits.

OBJ. 2
8.11(A)

McAllen's Oats *is the best tasting cereal on the planet!*

No other cereal tastes quite so delicious as McAllen's Oats. Many people enjoy eating our cereal with brown sugar and milk. Others eat it with berries, bananas, and other fruits.

Studies have shown that oatmeal is good for your health. People who eat oats each morning for two months can lower their cholesterol levels up to 20%!

So, what are you waiting for? Try McAllen's Oats today!

---

46 What scientific question could be asked about the claims made by the makers of McAllen's Oats?

A Does this cereal really have a delicious taste?
B Is this cereal more expensive to produce than other cereals?
C What evidence shows that this cereal lowers a person's cholesterol level?
D Can this cereal be eaten by someone when it is hot as well as cold?

OBJ. 1
8.3(B)

47 In the weather map below, a warm front is advancing towards the southeast.

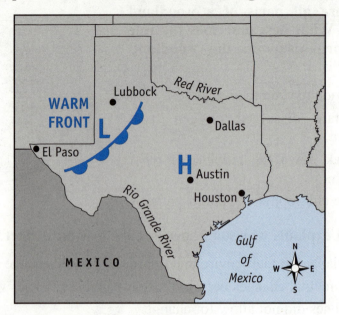

This map illustrates the interactions of the —

F   nitrogen and carbon cycles
G   solar and weather systems
H   ozone layer and air pollution
J   tectonic plates and the ocean system

OBJ. 5
8.10(B)

48 Which of the following is an important effect of tectonic plate movement?

A   Frequent tropical storms above lava flows
B   The creation of mountains where land plates collide
C   The recycling of organic compounds
D   The weathering and erosion of land forms where plates separate

OBJ. 5
8.14(A)

49 Which of the following lists elements from the Periodic Table in order from those with the fewest protons to those with most protons?

F   H, Na, Li
G   B, C, O
H   K, Na, Li
J   F, O, C

OBJ. 3
8.9(B)

50 Which of the following common substances is an element?

A   Oxygen gas (O_2)
B   Table salt (NaCl)
C   Carbon dioxide gas (CO_2)
D   Sugar ($C_{12}H_{22}O_{11}$)

OBJ. 3
7.17(C)

GLOSSARY

The blue bolded number following each entry indicates the page on which the term appears and is explained.

A

Atmosphere. Band of gases surrounding the Earth, made up mainly of nitrogen and oxygen. **[164]**

Atom. The smallest particle of an element. **[50]**

Atomic Mass. The number of protons and neutrons in an atom. **[52]**

Atomic Number. The number of protons in the nucleus of an atom. **[52]**

Axis. Imaginary line running through the center of the Earth around which the Earth rotates; this axis is tilted, causing the seasons. **[155]**

B–C

Bacteria. Unicellular microorganisms that lack a nucleus. **[105]**

Balance. An instrument for measuring mass. **[26]**

Carbon Cycle. Carbon dioxide is absorbed by plants, and released by animals, decomposers, and the burning of fossil fuels. **[165]**

Cell. The smallest basic unit of all living things. **[104]**

Cell Membrane. The membrane that holds cells together. **[105]**

Cell Theory. The theory that all living things are made of cells, which come from other cells. **[105]**

Cell Wall. Wall around the cell membrane of a plant cell, providing structure and support. **[106]**

Chemical Change. A change in a substance, resulting in a new substance. **[55]**

Chemical Equation. A chemical equation represents a chemical reaction; it always has the same number of atoms on each side of the arrow. **[57]**

Chemical Property. Ability of a substance to react with other substances. **[56]**

Chemical Reaction. A process that involves rearrangement of the atoms of one or more substances, such as combining oxygen and hydrogen to make water. **[56]**

Chromosome. A threadlike structure in the cell nucleus carrying genetic information in DNA. **[109]**

Circulatory System. System of human body made up of heart, blood vessels, and blood, which interacts with other systems to distribute oxygen and nutrients and to collect wastes. **[118]**

Climate. The average weather conditions of a place over a period of years. **[168]**

Community. All the different species of organisms living in a defined area. **[139]**

Compound. A substance formed from two or more elements chemically combined in fixed proportions. For example, water (H_2O), sugar ($C_{12}H_{22}O_{11}$) and salt (NaCl) are common compounds; a compound has different properties than the elements that make it up. **[54]**

Conservation of Mass. Matter cannot be destroyed by a physical or chemical change. **[56]**

Consumer. A living organism that lives by eating other organisms. **[43]**

Control Group. A group used as a basis of comparison for checking the results of an experiment; the control group is compared to the experimental group. **[24]**

Crust. The thin outer shell of Earth, made of rock. **[160]**

Cycle. A process in which the last step begins the process all over again, such as the water cycle, carbon cycle, rock cycle, or nitrogen cycle; materials are rearranged and reused in a cycle. **[164]**

Cytoplasm. The main material of a cell, inside the cell membrane but outside the nucleus; consists of fluid (cytosol) and structures. [104]

D

Data. Facts collected during an experiment or observed in nature. [28]

Decomposers. Organisms, such as bacteria and fungi, that break down dead organisms. [144]

Density. An object's mass divided by its volume (often measured in g/cm3). [65]

Dependent Variable. Something whose quantity changes as the result of a change made to the independent variable in an experiment. [23]

DNA. A complex molecule that contains the genetic code and transmits heredity in all living organisms. Watson and Crick developed the first DNA model. [109]

Dominant Trait. A trait, which if present in either gene of the parents, will appear in the individual; therefore, it dominates over recessive traits. [129]

E

Ecological Succession. Different organisms that arise in an area, creating a series of different ecosystems, one after the other, often after a catastrophic event. [143]

Ecosystem. An ecological system: a community of organisms and their environment. [138]

Electricity. A form of energy resulting from the flow of negatively charged particles (electrons). [92]

Electron. A subatomic particle with a negative electrical charge and almost no mass; it moves around the nucleus. [50]

Element. A substance with atoms of only one kind; all elements are shown on the Periodic Table. [54]

Endangered Species. A species close to extinction. [133]

Endocrine System. System of glands in the body producing hormones to regulate the body's activities; for example, thyroid hormone regulates the speed of bodily activities; insulin regulates blood sugar. [119]

Energy. The capacity for doing work; can be in such forms as kinetic energy, electrical energy, thermal energy and light. [90]

Equilibrium. A state in which different forces are balanced. [107]

Experiment. An attempt to test a hypothesis by gathering data under controlled conditions. [21]

Extinction. When the last member of a species dies. [133]

F

Feedback Mechanism. A way in which a system is able to monitor developments and adjust itself, such as the regulation of blood sugar in the human body by insulin. [108]

Folding. The bending of Earth's crust where tectonic plates collide, creating mountains. [161]

Food Chain or Food Web. An arrangement of the organisms of an ecosystem based on how they use each other as food sources; shows producers and different types of consumers. [144]

Force. An influence that results in a change of movement of a body in the direction it is applied: force = (mass)(acceleration); measured in newtons (N). [79]

Friction. The force from rubbing that resists the continued motion of two objects in contact. [79]

G

Gas. A state of matter without precise shape or volume, in which particles are spread apart. [64]

Gene. The part of a chromosome that determines a specific trait. [128]

Gravity. A force of attraction between two objects in proportion to their masses, and decreased by a greater distance between them; gravity affects the movements of the moon and planets. [153]

Groundwater. Water that soaks into the ground and collects above layers of solid rock; it forms part of the watershed of an area. [165]

H–I

Heredity. Characteristics inherited from one's ancestors, based on genes. [131]

Hypothesis. An educated guess that attempts to answer a scientific question. [22]

Igneous Rock. Rock made from cooled magma. [164]

Independent Variable. A variable that a scientist changes to find out how this change affects other variables in an experiment. [23]

Inherited Traits. Traits passed on from the parents to their children. [128]

J–L

Kinetic Energy. The energy of motion; for example a falling object has kinetic energy. [91]

Land Subsidence. A weak area in the Earth's crust that collapses inward. [163]

Lever. A simple machine that allows a smaller force exerted over a larger distance to balance a larger force over a shorter distance. [81]

M

Magma. Molten rock material within the Earth; becomes known as *lava* at the surface. [162]

Mass. The amount of matter an object has; it is proportional to its weight; usually measured in grams or kilograms. [50]

Matter. Anything that occupies space and has mass. [49]

Metamorphic Rock. Rock, such as marble or slate, made from igneous or sedimentary rock that has been changed by heat and pressure under the Earth's surface. [164]

Metric System. A system of weights and measurements based on the meter, gram and liter. [28]

Mixture. Two or more substances mixed together but not chemically combined, such as salt water or salt and sand; the substances can be separated by physical means. [70]

Molecule. A group of atoms joined by sharing electrons. [54]

Mutation. A hereditary change in a gene caused by chance or by environmental conditions. [132]

N–O

Natural Selection. The theory that individuals having characteristics that aid in survival will have more offspring, so that the proportion of such individuals in a species will gradually increase. [132]

Neutron. An uncharged particle found in the nucleus of an atom with a mass equal to a proton. [50]

Nucleus. (1) The center of an atom consisting of protons and neutrons. (2) The part of many living cells surrounded by a nuclear membrane and storing the cell's DNA. [50]

Observation. Something an experimenter sees, hears, smells, feels, or tastes. [19]

P–Q

Particle. Any of the basic units of matter (*molecule, atom, proton,* or *electron*). [50]

Periodic Table. A table showing all the known elements, arranged into periods and groups based on their common properties and electron arrangement. [67]

Phases of the Moon. Changes in the shape of the moon as observed from the Earth. [156]

Photosynthesis. The process by which green plants use light energy to convert carbon dioxide and water into carbohydrates, storing energy and releasing oxygen. [108]

Physical Properties. A property of a substance that may change without changing its chemical makeup: color, hardness, odor, state (gas, liquid, solid), boiling and freezing points. [62]

Plant Cells. Have cell walls; conduct photosynthesis; have large vacuoles for waste. [106]

Potential Energy. The energy stored in an object based on its location or situation, such as chemical energy, or the energy stored in a compressed spring. [91]

Predator. An animal that lives by eating prey. [142]

Producer. An organism that produces its own food, such as a green plant. [143]

Proton. A subatomic particle with a positive electrical charge and an atomic mass of one. [50]

Pulley. A simple machine of a rope or cord around a wheel; two linked pulleys allow a heavier weight to be lifted a shorter distance by less force applied over a longer distance. [82]

Punnett Square. A diagram used to predict the probability that offspring will inherit particular traits; capital letters are used for dominant traits, while small letters are for recessive traits. [130]

R–S

Recessive Trait. A genetic trait that appears only if shared by both parents' genes; otherwise, it is masked by the dominant trait but may reappear in later generations. [129]

Rock Cycle. Recycling of material between igneous, sedimentary, and metamorphic rocks. [164]

Scientific Investigation. The method scientists use to investigate the natural world and conduct experiments by asking a question, forming a hypothesis, making a prediction, testing the hypothesis, gathering data, drawing conclusions, and communicating results. [33]

Scientific Knowledge. Knowledge based on careful observation, experimentation, and the testing of theories attempting to explain the natural world. [21]

Sedimentary Rock. Rock created in layers by the accumulation and compression of sand, mud, and other sediment. [164]

Solid. A state of matter with definite density and shape, in which molecules are closely packed. [64]

Solution. A mixture in which one substance is evenly dispersed in another. [70]

Sound Waves. Mechanical energy transmitted by waves in the air or some other medium, which we hear as sounds. [95]

Species. A group of organisms of similar structure, capable of mating and producing offspring. [132]

Specific Heat. Amount of heat energy needed to raise 1 gram of a substance by 1 degree Celsius. [65]

Speed. The average distance an object moves in a given unit of time, often measured in m/s. [77]

Star. A giant ball of gas in space, like the sun, that produces energy through nuclear fusion. [152]

Substance. Any element or compound. [56]

System. A group of parts that work together to achieve a function or task that they could not achieve alone. [116]

T–U

Technology. The application of scientific knowledge to develop tools, materials, and systems that help humans meet their needs. [42]

Temperature. The average kinetic motion of particles in an object, measured in degrees. [65]

Theory. A possible explanation of observations and data, which can be tested and revised. [20]

Trait. An inherited characteristic. [127]

Turgor Pressure. Pressure of a plant cell against its cell walls, based on its water content. [108]

V–Z

Variable. A quantity that can change. [23]

Water Cycle. Water passes from vapor in the atmosphere through precipitation to land or water surfaces, and then back into the atmosphere through evaporation and transpiration. [164]

MIDDLE SCHOOL TAKS OBJECTIVES IN SCIENCE

The following pages list all of Texas' science objectives with their related student expectations and grade level expectations. These science standards have been referred to throughout this book. In particular, every question has been identified by its specific objective, student expectation and grade level. These are identified by their objective — such as "**Obj. 2**" for Living Systems and the Environment, and by the grade and student expectation — "**6.10**" for its grade level and "B" for the student expectation.

**OBJ. 2
6.10 (B)**

OBJECTIVE 1: NATURE OF SCIENCE

(6.1, 7.1, 8.1)   **Scientific processes.** The student conducts field and laboratory investigations using safe, environmentally appropriate, and ethical practices. The student is expected to

(A) demonstrate safe practices during field and laboratory investigations.

(6.2, 7.2, 8.2)   **Scientific processes.** The student uses scientific inquiry methods during field and laboratory investigations. The student is expected to

(A) plan and implement investigative procedures including asking questions, formulating testable hypotheses, and selecting and using equipment and technology;

(B) collect data by observing and measuring;

(C) organize, analyze, evaluate, make inferences, and predict trends from direct and indirect evidence (7.2, 8.2);

(D) communicate valid conclusions; and

(E) construct graphs, tables, maps, and charts using tools to organize, examine, and evaluate data.

(6.3, 7.3, 8.3)   **Scientific processes.** The student uses critical thinking and scientific problem solving to make informed decisions. The student is expected to

(A) analyze, review, and critique scientific explanations, including hypotheses and theories, as to their strengths and weaknesses using scientific evidence and information;

(B) draw inferences based on data [related to promotional materials] for products and services; and

(C) represent the natural world using models and identify their limitations.

(6.4, 7.4, 8.4)   **Scientific processes.** The student knows how to use a variety of tools and methods to conduct science inquiry. The student is expected to

(A) collect, record, and analyze information using tools including beakers, petri dishes, meter sticks, graduated cylinders, weather instruments, hot plates, dissecting equipment, test tubes, safety goggles, spring scales, balances, microscopes, telescopes, thermometers, calculators, field equipment, computers, computer probes, water test kits, and timing devices (8.4); and

(B) extrapolate from collected information to make predictions (8.4).

OBJECTIVE 2: LIVING SYSTEMS AND THE ENVIRONMENT

(6.5) **Science concepts**. The student knows that systems may combine with other systems to form a larger system. The student is expected to

(B) describe how the properties of a system are different from the properties of its parts.

(6.10, 7.9) **Science concepts**. The student knows the relationship between structure and function in living systems. The student is expected to

(B) determine that all organisms are composed of cells that carry on functions to sustain life (6.10); and

(C) identify how structure complements function at different levels of organization including organs, organ systems, organisms, and populations (6.10).

(6.8, 7.8, 8.10) **Science concepts**. The student knows that complex interactions occur between matter and energy. The student is expected to

(B) identify that radiant energy from the Sun is transferred into chemical energy through the process of photosynthesis (7.8).

(7.12) **Science concepts**. The student knows that there is a relationship between organisms and the environment. The student is expected to

(B) observe and describe how organisms including producers, consumers, and decomposers live together in an environment and use existing resources;

(C) describe how different environments support different varieties of organisms; and

(D) observe and describe the role of ecological succession in ecosystems.

(8.6) **Science concepts**. The student knows that interdependence occurs among living systems. The student is expected to

(A) describe interactions among systems in the human organism;

(B) identify feedback mechanisms that maintain equilibrium of systems such as body temperature, turgor pressure, and chemical reactions; and

(C) describe interactions within ecosystems.

(6.11, 7.10, 8.11) **Science concepts**. The student knows that traits of species can change through generations and that the instructions for traits are contained in the genetic material of the organisms. The student is expected to

(A) identify that change in environmental conditions can affect the survival of individuals and of species (8.11);

(B) distinguish between inherited traits and other characteristics that result from interactions with the environment (8.11); and

(C) make predictions about possible outcomes of various genetic combinations of inherited characteristics (8.11).

OBJECTIVE 3: STRUCTURES AND BODIES OF MATTER

(6.7, 7.7, 8.9)   **Science concepts.** The student knows that substances have physical and chemical properties. The student is expected to

- (B) classify substances by their physical and chemical properties (6.7);
- (C) recognize that compounds are composed of elements (7.7);
- (A) demonstrate that substances may react chemically to form new substances (8.9);
- (B) interpret information on the periodic table to understand that properties are used to group elements (8.9); and
- (C) recognize the importance of formulas and equations to express what happens in a chemical reaction (8.9).

(8.8)   **Science concepts.** The student knows that matter is composed of atoms. The student is expected to

- (A) describe the structure and parts of an atom; and
- (B) identify the properties of an atom including mass and electrical charge.

(6.8, 7.8, 8.10)   **Science concepts.** The student knows that complex interactions occur between matter and energy. The student is expected to

- (A) illustrate interactions between matter and energy including specific heat (8.10).

OBJECTIVE 4: MOTION, FORCES, AND ENERGY

(6.9)   **Science concepts.** The student knows that obtaining, transforming, and distributing energy affects the environment. The student is expected to

- (A) identify energy transformations occurring during the production of energy for human use such as electrical energy to heat energy or heat energy to electrical energy.

(6.8, 7.8, 8.10)   **Science concepts.** The student knows that complex interactions occur between matter and energy. The student is expected to

- (A) illustrate examples of potential and kinetic energy in everyday life such as objects at rest, movement of geologic faults, and falling water (7.8).

(6.6, 7.6, 8.7)   **Science concepts.** The student knows that there is a relationship between force and motion. The student is expected to

- (B) demonstrate that changes in motion can be measured and graphically represented (6.6);
- (A) demonstrate basic relationships between force and motion using simple machines including pulleys and levers (7.6);
- (C) relate forces to basic processes in living organisms including the flow of blood and the emergence of seedlings (7.6);
- (A) demonstrate how unbalanced forces cause changes in the speed or direction of an object's motion (8.7); and
- (B) recognize that waves are generated and can travel through different media (8.7).

OBJECTIVE 5: EARTH AND SPACE SYSTEMS

(6.14) **Science concepts.** The student knows the structures and functions of Earth systems. The student is expected to

(B) identify relationships between groundwater and surface water in a watershed.

(6.13, 7.13) **Science concepts.** The student knows components of our solar system. The student is expected to

(A) identify and illustrate how the tilt of the Earth on its axis as it rotates and revolves around the Sun causes changes in seasons and the length of a day (7.13); and

(B) relate the Earth's movement and the moon's orbit to the observed cyclical phases of the moon (7.13).

(6.8, 7.8, 8.10) **Science concepts.** The student knows that complex interactions occur between matter and energy. The student is expected to

(B) explain and illustrate the interactions between matter and energy in the water cycle and in the decay of biomass such as in a compost bin (6.8); and

(B) describe interactions among solar, weather, and ocean systems (8.10).

(8.12) **Science concepts.** The student knows that cycles exist in Earth systems. The student is expected to

(A) analyze and predict the sequence of events in the lunar and rock cycles; and

(C) predict the results of modifying the Earth's nitrogen, water, and carbon cycles.

(8.13) **Science concepts.** The student knows characteristics of the universe. The student is expected to

(A) describe characteristics of the universe such as stars and galaxies.

(7.14, 8.14) **Science concepts.** The student knows that natural events and human activity can alter Earth systems. The student is expected to

(A) describe and predict the impact of different catastrophic events on the Earth (7.14);

(B) analyze effects of regional erosional deposition and weathering (7.14);

(C) make inferences and draw conclusions about effects of human activity on Earth's renewable, non-renewable, and inexhaustible resources (7.14);

(A) predict land features resulting from gradual changes such as mountain building, beach erosion, land subsidence, [and continental drift] (8.14); TAKS will assess students' understanding of plate tectonics. The theory of plate tectonics is the most current and accepted theory of plate movement.

(B) analyze how natural or human events may have contributed to the extinction of some species (8.14); and

(C) describe how human activities have modified soil, water, and air quality (8.14).

INDEX

A

Acceleration, 80
Acid rain, 170, 173
Adaptation, 132
Adaptive radiation, 134
Aquatic ecosystem, 1398
Aristotelian View of universe, 40
Arteries, 115
Atmosphere, 164, 165, 195
Atomic mass, 50, 52–53, 58, 195
Atomic number, 52, 53, 58, 71, 195
Atomic symbol, 52, 53, 58
Atoms, 49, 50–52, 195
Average, computing an, 30
Axis, 155, 195

B

Bacteria, 105, 109, 195
Balance, 26, 195
Barometer, 25
Biodiversity, 171
Black hole, 152

C

Capillaries, 115
Carbon cycle, 165–166, 195
Carnivores, 143
Cell, 104, 195
   division, 109, 111
      membrane, 104, 105, 195
      theory, 105, 111, 195
      wall, 106, 195
Cellular respiration, 106, 109, 110
Chemical energy, 92
Chemical equation, 57, 58, 195
Chemical reaction, 55–56, 195
Chloroplasts, 106
Chromosome, 109, 195
Circulatory system, 118, 120, 123, 195
Climate, 168, 195
Comets, 154
Community, 139, 145, 195
Competition, 142
Compound, 54, 56, 58, 70, 195
Connective tissue, 115
Conservation of energy, 95
Conservation of mass, 56, 58, 195
Consumer, 138, 143, 144, 145, 195

Continental plates, 160
Contour maps, 17
Control group, 24, 34, 195
Controlling variable, 23, 34
Cooperation, 142
Copernican Theory, 40
Crick, Francis, 131
Crust, Earth's, 160, 195
Cycle, 164, 195
Cytoplasm, 104, 111, 195

D

Darwin, Charles, 134
Data, 28, 195
Decomposers, 138, 144, 145, 196
Density, 62, 65, 195
Dependent variable, 23, 34, 196
Deserts, 140
Digestive system, 117, 120, 123
Dissecting equipment, 25
Distance, 77, 81
D.N.A., 104, 109, 110, 129, 131, 134, 135, 196

E

Earth,
   cycles, 164–166
   movement of, 154–155
   processes, 169–170
Earthquakes, 162
Ecological succession, 138, 143, 145, 196
Ecology, 138–144
Ecosystem, 138, 145, 196
Electricity, 92, 196
Electromagnetic
   radiation, 99
   spectrum, 96
   waves, 95, 96, 98
Electron, 50, 58, 196
Electron cloud, 50
Element, 49, 54, 56, 58, 196
Endangered species, 173
Endocrine system, 119, 120, 123, 196
Energy, 90–95, 98, 196
Energy levels, 50
Energy, transformation of, 93–95
Environmental change, 135
Equilibrium, 107, 111, 123, 142, 196
Erosion, 163, 172
Excretory system, 119, 120, 123

Experiment, 21, 196
Experimental design, 23–24
Extinction, 133, 171, 173, 196

F

Feedback mechanism, 108, 121–122, 123, 196
Folding, 161, 196
Food chain, 144
Folding, 161, 196
Force, 79, 81, 84, 85 Folding, 161, 196
   mechanical, 84, 85
   balanced, 77, 85
   unbalanced, 77, 85
Franklin, Rosalind, 131
Frequency, 30, 97
Friction, 79, 196

G

Galaxy, 151, 153, 157
Gamma rays, 97
Gas, 64, 196
Gear, 83
Gene, 128, 135, 196
Global warming, 170, 173
Graduated cylinder, 25
Grassland, 141
Gravity, role of, 153–154, 196
Greenhouse effect, 170
Groundwater, 165, 196
Groups, experimental, 67, 72

H

Heat energy, 91
Herbivore, 143
Heredity, 127–132
Hormones, 119
Humidity meter, 25
Hurricane, 169
Hypothesis, 22, 34, 197

I-J-K

Igneous rock, 164, 197
Inclined plane, 82
Independent variable, 23, 34, 197
Insulin, 122
Integumentary system, 119, 120, 123
Interdependent, 138, 145
Internal combustion engine, 94
Kinetic energy, 90, 91, 98, 197

L

Laboratory equipment, 25–26, 34
Land subsidence, 163, 172, 197
Lava, 162
Lever, 81, 85, 197
Lunar phases, 156, 157, 197
Lysosomes, 106

M

Machines, 81, 85
   power-driven, 94
Magma, 162, 197
Mass, 50, 197
Material Safety Data Sheets, 27
Matter, 49, 197
   states of, 63–64, 71
Mechanical wave, 95, 99
Mendel, Gregor, 128–129
Mendeleev, Dmitri, 66
Meniscus, 28
Metalloid, 68–69, 72
Metals, 68–69, 72
Metamorphic rock, 164, 197
Meter stick, 25
Metric System, 28, 197
Microwave, 97
Mid Atlantic Ridge, 162
Mixture, 62, 70, 72, 197
Models, 21, 34, 38–39
Molecule, 54, 58, 197
Moon, 154
Motion, 77, 79
Multicellular organism, 114
Muscle tissue, 114
Muscular system, 117, 120, 123
Mutation, 132, 197

N

Natural selection, 132–133, 134, 135, 197
Nebula, 152
Nervous system, 118, 120, 123
Nervous tissue, 115
Neutron, 50, 53, 58, 197
Newton, Sir Isaac, 41, 44, 79
Newtons, 80, 84
Nitrogen cycle, 165–167
Noble gases, 69
Nonmetals, 68–69, 72
Non-renewable resource, 171, 173
Nuclear energy, 92, 99
Nuclear fusion, 152
Nucleus, 50, 105, 109, 111, 197

O-P-Q

Omnivore, 143
Organ system, 114, 115, 123
Organ, 114, 115, 123
Ozone layer, 170, 173
Parasites, 142
Pasteur, Louis, 41, 44
Percentage, 30
Periodic Table of the Elements, 54, 62, 66–69, 71, 72, 197
Periods, 67
Pesticides, 170
Photosynthesis, 104, 106, 108, 110, 111, 197
Planets, movement of, 154
Plant cell, 106, 198
Pollution, 170
Population, 132, 135, 139, 145
Potential energy, 90, 91, 98, 198
Precipitation, 165
Predator, 142, 198
Producer, 138, 143, 144, 145, 198
Proton, 50, 53, 58, 198
Pulley, 82–83, 85, 198
Punnet square, 130–131, 134, 135, 198

R

Range, 30
Red giant, 152
Renewable resources, 171
Reproductive system, 120, 123
Respiratory system, 118, 120, 123
Revolution, 155
Ribosomes, 106
Rock cycle, 164, 198
Rotation, 155, 157

S

Safety equipment, 27
Scientific inquiry, 19, 33
Scientific investigation, 21–32, 33, 198
Scientific theory, 20, 33, 38, 41
Sedimentary rock, 164, 198
Seismic wave, 162
Semiconductor, 68
Sexual reproduction, 110
Skeletal system, 117, 120, 123
Skin tissue, 115
Solar system, 151, 153
Solid, 64, 198
Solution, 70–71, 72, 198
Species, 132, 135
Specific heat, 65, 71
Speed, 77, 85, 198
Spring scale, 25
Star, 151, 152, 157, 198
Substance, 56, 198
Sun, 153, 157
System, 114, 115, 123, 198

T

Technology, 42–43, 198
Tectonic plates, 160, 172
Telescope, 25
Temperate Forest, 140
Temperature, 65, 198
Theory, 20, 198
Theory of Evolution, 133
Thermometer, 25
Tilt, Earth's 155, 157
Timing device, 26
Tissue, 114, 123
Tornado, 169
Traits
   dominant, 127, 129, 134, 135, 195
   inherited, 128
   recessive, 127, 129, 134, 135, 198
Transpiration, 165
Trend, 29–30
Triple beam balance, 26
Tropical rain forests, 140
Tundras, 141
Turgor pressure, 108, 111, 198

U-V

Ultraviolet light, 97
Unicellular organism, 114
Universe, 151, 157
Vacuoles, 106
Variable, 23, 198
Veins, 115
Visible light, 97
Volcano, 162

W-X-Y-Z

Warm front, 16
Water cycle, 164, 198
Watershed, 165
Water-test kit, 26
Watson, James, 131
Wave, 90
Weather
   instrument, 25
   map, 16
   pattern, 168
Weathering, 163, 172
Wedge, 82
Weight, 50
Wheels, 83
Wind, 169
X-ray, 97